Don Gree[illegible]

THE BUG BOOK

harmless insect controls

HELEN and JOHN PHILBRICK

This informal presentation has been assembled in response to urgent requests of many of our gardening friends for simple ways to control backyard garden pests without using toxic materials.

Illustrated by
CATHY BAKER

GARDEN WAY PUBLISHING · CHARLOTTE · VERMONT

The Authors assume responsibility for statements made: that they are either personal experiences of the Authors or statements made by other working gardeners. The methods used have as far as possible been in accord with the Bio-Dynamic Method but no statements herein are to be considered in any way official statements of the Bio-Dynamic Farming and Gardening Association, Inc.

Printed in the United States of America
Eighth Printing, June, 1979

Library of Congress Catalog Card No. 74-75470
ISBN: 0-88266-027-6

table of contents

To the many back yard gardeners who devote their waking hours to maintaining, in integrity, a personal link between themselves and the Created Universe within which we are presently living.

"... to try to synthesize nature's kingdom"

Great credit must always go to the research persons who have given their careers to the study of insects and birds and mammals with their respective relationships, each to the other. There have often been economic considerations prompting such research. In fact, the economic element has carried so much weight that many times in the history of every country, bureaus have been established to further these studies. Government bulletins and the encyclopedias are filled with scholarly analyses reported with scientific facts and uncontestable accuracy in every detail. Much gratitude is due these men and women for the great wealth of information they have given to the world concerning the small creatures surrounding us on all sides.

Now that we know all the minutiae about the least gnat,—how it reproduces, how its salivary glands function, even how characteristics of its salivary glands are passed down to successive generations, it seems as though we might start to learn about the whole gnat. Not only could we learn more about the whole living gnat, but we might use an equal amount of analysis and concentration to learn about the gnat in relation to other gnats and the gnats in relation to the other creatures in their environment.

After the naturalists have dissected thousands of birds to count the insects in their stomachs, as is conscientiously recorded in scientific natural history, it is encouraging to see such groups as the Audobon Society emphasizing the study of living birds and their habits in real life situations.

insect control in our own back yard

It is not our intention in this small book to make an exhaustive treatise on all the scientific work that is being done on biological controls and insect sterilization by radiation. We are content to leave that to the scientists and the white-coated technicians who like to spend their days peering through microscopes.

Our first interest is in the small productive piece of land on which we live with myriad other living creatures. First of all we want a productive vegetable garden which provides enough food for our use, with perhaps a few flowers for our esthetic sense. We also have about a dozen fruit trees, some old and some younger, and we have special concern that they should provide fruit enough for the summer and fall, with surplus for freezing and canning and juice-making. A flock of hens provides some livestock, and over all the honey bees fly hither and yon adding their own special vitalities to herbs and animals. The wild birds form a link between the domesticated hens who must live on the ground and the completely wild woods dwellers who seldom come even to our bird feeder. All winter we have literally hundreds of birds (150 Evening Grosbeaks by actual count); and the summer population which works for us in garden, lawns and fruit trees is impossible to number.

Along with the wild birds work the untameable insects, in utter freedom to come and go as they please. We will not poison them (even though we have listed a few strong measures among these formulae for the use by more timid people who wince when an insect comes near them.) We are very much aware that there are many more repellents and baits and controls than are mentioned here. These in this book are primarily the materials and substances which we have tried ourselves and have found effective in our life's study of how to live at peace with insects.

For as far back as I can remember, I have viewed all insects with an innate feeling of curiosity, affection and the realization that they are an integral part of our natural ecology. If man by some dry intellectual scientific process should succeed in sterilizing all the male insects by radiation, we have some reason to believe that the resulting vacuum in the world of nature would very soon result in man's inability to perpetuate himself on this planet. Although we have no claim to knowledge except that gained by experience, we prefer to use the gentler methods of insect control instead of those aimed at total extinction.

The products and the nutritional values of the backyard garden may be assayed in a laboratory. There should be no temptation to compromise or to accept second best. One has to be utterly honest, to make painstakingly accurate observations of what is happening among and between growing things. For instance, it is not enough to theorize that the wild birds are helpful because they keep busy eating insects. One must make a closer study in the living situation to know from firsthand observation that they eat insects and exactly what insects they eat. Similarly it is not enough to theorize that vegetables and fruit grown in the home garden have superior qualities. Chromatograms made in the laboratory give proof of the quality of the life process present in the vegetables being tested. But in the long run, even the laboratory analysis is not an integral part of everyday living. The final test of the goodness of home grown food is the health and good humor of the people who eat it. This is another firsthand, personal observation which seems to carry some weight with our family, perhaps because we are the ones who enjoy the home grown produce!

From our everyday experience then, we can say that our three acres, plus the neighboring woods and meadows, plus the manure from our neighbor's cowbarn, all surrounding our garden and fruit trees, have provided for us excellent nutrition and a state of good health for many happy years. The wild birds and insects fit into the pattern without molestation except when some one detail gets out of hand. For the past five years, for instance, we have not had to use even rotenone for insect control. By adequate fertilization by Compost (made according to the rules,) and by specific treatment of the soil, and by observing the companion plants which are mutually helpful and which sometimes repel certain insects, and by occasional handpicking, we have not had to resort to any stronger methods of insect control. Strong healthy plants growing in a fertile soil seem to be able to resist insect damage. Who knows but that in future conquests of the Space Age man may come to understand how and why insects are a necessity for well-balanced fertility!

We have learned these truths of nature: how the birds have chosen the higher altitudes above the earth. They belong in general to the level which includes the trees. In similar pattern the insects belong in general to the lower level of the atmosphere where the plants dwell. Because the plants and the trees are rooted in the earth and cannot move about, they require the services of some more mobile creatures to fetch and carry some of the earth and cosmic forces, about which man as yet knows but little. The insects are also the link between plant roots and the soil and the thousands of soil organisms which are still unsearched. Both birds and insects have developed with entirely free modes of locomotion and activity. The bird lives in a tree and brings down into the twigs the wonders he beholds from the heights of the skies. The butterfly brings to the earthbound flowerbed the freedom she has experienced as she fluttered free over the meadow. Both birds and insects help to complete the growth of tree and plant as they dwell fastened in the earth but stretching up toward the sun and the stars into the sky. Insects, furthermore, because of their metamorphoses, encompass in their own development from worm to pupa to butterfly, the total experience of living first on the ground, and then above ground and then in the air above the earth. For water insects and the tiny soil bacteria which belong to the insect sphere, there are still other purposes of bringing together soil and water and growing plant. The bio-chemists of the future will learn many facts about all these creatures and their relationships. But no matter how much is learned in the scientific world, the place where it matters most to you and to me is in our own back yards, where these living creatures in abundant vitality are free to coexist with each other and with appreciative men and women and children.

the bio-dynamic method

The working principles of the Bio-dynamic agricultural method were established in 1924 by Dr. Rudolf Steiner at a farmers' conference held at Koberwitz in Germany as an attempt to renew their older farming practices.

With enough foresight those farmers could already see that increased use of chemical fertilizers and pesticides would in time deplete soil life to the extent that soil fertility as such would completely collapse. Dr. Steiner spoke from his childhood observations of organic practices in peasant agriculture in southern Europe (about one hundred years ago now.) To his observations Dr. Steiner added his own remarkable (seemingly clairvoyant) insight and his ability to recognize forces and influences in nature that are invisible to most of us, until they are pointed out. As soon as we learn to recognize such forces, we can see their influences at work wherever there are living plants, animals or people.

Today Bio-dynamic farmers and gardeners make an intensive study of all available facts related to earth, plants, insects, birds, fish and animals. In addition they make a scientific study of the sun, moon, fixed stars and the planets. They apply their scientific knowledge of cosmic happenings to the actual practices of farming, to produce the best possible crops.

Modern scientific tests and experiments have indicated the validity of many techniques which came down from past practices and which may seem archaic. There is abundant scientific research to indicate that forces and influences from outer space can actually be directed into earth agriculture by the use of Bio-dynamic preparations in the compost pile. For this, see the article "Biodynamics: What it is and What it is Not" by E. E. Pfeiffer, which explains tests by International Harvester Laboratories in analyzing compost before and after treatment with BD preparations (available from Biochemical Research Laboratory, Spring Valley, N.Y. 10977.)

Insect pests can be kept to a minimum by the presence of healthy plants. Certain companion plant combinations are indicated specifically to keep down certain insects. Other homemade insect controls are time-tested, and are as numerous as the gardeners who use them. This book contains as many of these controls as we know to be effective.

some general principles of insect control from long ago

quoted from an 1824 ENCYCLOPEDIA OF GARDENING

compiled by J. C. Loudon

This Encyclopedia gives in minute detail the aims and the description of gardeners' programs which resulted in the extensive gardens and estates of Europe which were at their height in Loudon's time. The head gardeners in those days possessed a botanical knowledge and a horticultural skill unsurpassed in any era. The majority of their plants and most of the insects he describes are identical with ours in the Twentieth Century—even to his treatise on Purple Cauliflower.

"*The preventive operations* are those of the best culture in the most extensive sense of the term, including what relates to choice of seed or plant, soil, situation, and climate. If these are carefully attended to, it will seldom happen that any species of insect will exist in the garden to an injurious degree. But some parts of culture such as climate are often beyond our control; as for example when a very dry spring and east winds prevail, in which case many insects increase, or rather their larvae are hatched and reared under such favorable circumstances that few of them die. They become strong in proportion to the plants on which they live, which, in consequence of the dry weather, become weak. In such a case as this, or its reverse, that of a series of cold moist weather, the gardener cannot apply good culture to the plants in the open air, and therefore cannot prevent the increase of insects.

"*The Palliative operations* are various: Artificial bad weather will annoy every description of organized being and especially insects. Excessive waterings, stormy applications of water with a syringe, violent wind; these and similar operations will materially injure and annoy insects both in their common functions and in the work of generation, hatching and rearing. Insects may be further annoyed by throwing on them acrid waters or powders, as tobacco-water, lime-water, powdered quick-lime, soot, ashes, barley-awns, etc., etc. The smell of tar is particularly offensive to various moths and butterflies; and it is said, if a little of it is placed under plants, or if they are watered with tar-water, these insects will not lay their eggs on them. It is also said that if shreds of flannel are hung on trees or plants, moths and butterflies will lay their eggs on the shreds in preference to the leaves of the plant. The effect of the fumes of tobacco, sulphur, urine etc. are well known. Saline substances mixed with water are injurious to most insects with tender skins, as the worm and slug. Hot water where it can be applied without injuring vegetation, is equally, if not more powerfully, injurious. Water heated to 120 or 130 degrees will not injure plants whose leaves are fully expanded and in some degree hardened, and water at 200 degrees or upwards may be poured over leafless plants. There are various other ways in which insects may be annoyed, and often in part destroyed, which will be pointed out in treating of the plants which particular species inhabit. The effects of insects may also be palliated on one species of plant, by presenting to them another which they prefer; thus wasps are said to prefer carrots, the berries of the yew and the honey of the hoya, to grapes; honey or sugared water to ripe fruit, and so on. One insect or animal may also be set to eat another, as ducks for slugs and worms, turkeys for the same purpose, and caterpillars and ants for aphides, and so on.

"*The operations for the utter removal or destruction of insects* are few, and chiefly that of handpicking, or otherwise removing or killing by manual operations with a brush, sponge or net. Destruction by handpicking should, if possible, commence with the parent insect in its fly, or perfect state before it has deposited its ova. Thus the gathering of moths, butterflies, and large wasps may save the gathering afterwards of thousands of caterpillars and the drowning of hundreds of wasps as preventing weeds from seeding in a garden will soon eradicate them altogether. It is no small proof of the advantages of a knowledge of natural history to gardeners, and also of the progress of knowledge among this ingenious and useful class of artisans, that a practical gardener has actually practiced for several years the catching of moths, to prevent them from laying their eggs on his trees.

"P. Musgrove, gardener, at May-field near Edinburgh, has almost completely cleared his trees of caterpillars by the following mode:

'I examine,' he says 'the trees I wish to clear in the beginning of June, that being the time the moths begin to leave the chrysalis state. When I find one of those of a dark color, I am aware the insect will make its appearance in the course of a few days. That chrysalis I examine daily until the insect comes out; and although I do not see the insect emerging from the shell, yet I am sure to find it in the neighborhood of the covering which it has left, exhausted with fatigue in consequence of the exertion in extricating itself from its confinement. At first I put a few of the chrysalids into paper bags, which gave me an opportunity of examining them minutely. I also watched some of the chrysalids of the bore-worm, which causes gooseberries to fall off in great quantities by boring into the berry, and I found that fly to be of the same class with those which infest the apple, pear and cherry trees. I was also able to prove decidedly that the females come into existence full of the rudiments of eggs, which I found by dissecting several of them and examining the ovarium. I also found, by carefully noticing every insect which I caught, that the greater number were females.' "

Some of these observations sound like 20th Century science. Imagine such a wealth of factual material being buried in a book since 1824!

"Having made himself completely acquainted with the enemy with which he had to contend, he continued his labors: 'Going over a number of wall trees which I fixed upon for the experiment, with a branch of a willow-tree in my hand, with which I switched the leaves and branches, for it is amongst the leaves and branches of the trees the insect secretes itself; but in order that it may be done with more expedition and success, I would recommend a birch-besom [broom] to be used in preference. There should be two persons, one to go over the leaves and branches of the trees, in order to make the insect leave its retreat and one with a net attached to a pole to catch the fly or to destroy it if it should alight on the ground, as it will be apt to do if the day is clear and sunny, for these insects cannot bear the bright rays of the sun which is the cause of their remaining amongst the leaves during the day; but should the day be dull, the net will be highly necessary to catch the insect, as it will then likely fly to some distance before it alights. This operation must be continued until all the insects are destroyed; but it is not needful that it should be performed every day, but every other day, as the insects are some days from the chrysalid state before they are ready to deposit their ova, which is done during the night.

" 'The method followed with standard trees is as follows: the time for going over them is generally two or three weeks later than the wall trees. It is a singular fact, that the insect keeps pace with the leafing of the tree. With the standards nothing will be required but the net, as the branches can be gently shaken, which is sufficient to cause the fly to leave its nesting place; but as it might be the means of bringing too many down at one time, if the tree was shaken all at once care must be taken to shake the branches one by one. Where the trees are lofty, a pole with a hook attached to the end may be used.

" 'The net used is made of strong black gauze, that color being best for the purpose. It is a yard and a half in circumference, a foot deep, and attached to a whalebone rim. The handle is made of common wood, about a yard and a half long. With regard to the manner in which it should be used, all I have to say is, that I kept the net in my right hand; and the moment an insect was driven from its place, I swung the net in the direction opposite to that in which it flew. If I missed in the first attempt, the second generally succeeded.' "

If the picture of this leisurely net-swinging gentleman pursuing moths among his apple trees causes you to chuckle, compare it for a moment with our gas-masked, plastic-suited technician on his tractor with a spray rig, pursuing moths among his apple trees in our scientific Twentieth Century!

Loudon goes on to mention some of the background a gardener should acquire, because there were no Agricultural Colleges in those "unenlightened" days:

"The young gardener should carefully and assiduously study the nature, names, and classifications of insects and make himself acquainted with all the species he can pick up, either in gardens, houses or fields. Besides being of material use in his profession, he will find it a never failing source of interest and enjoyment, at least equally so with the study of botany.

"The operations for the cure of accidents to trees are chiefly of cutting off injured parts, supporting and coating over. Amputation must be performed with suitable instruments and so as to leave a smooth section calculated to throw off the water.

"For coating over wounds in trees and plants: the usual application is now clay and loam made so thin as to be laid on with a brush and two or three coats may be given. On large wounds paint, or putty and paint may be used; and in the case of deep hollow wounds, the part may be filled with putty, or putty and small stones, for the sake of saving the former, and then made smooth and well painted.

"The operations for curing diseases are few, besides those for the cure of accidents. Washes are applied by the sponge, brush and syringe or watering pot, for filth, mildew and blight."

how to identify the insects in your garden

We have a friend who eagerly "got him a garden" and developed an overpowering enthusiasm for all the little live creatures he had never before noticed. His family joined him with equal ardor and everything went well until relatives came to visit. The latter continued in their accustomed way, killing every "bug" they met. Our friend put up such a howl that the killing ceased abruptly. "Now" he says, "they run into the house with every bug they find, and ask 'Is this a good 'un or a bad 'un? Should we kill it, or do you want us to put it back in your garden?' "

Most of the common insects are easy to find in a popular handbook like the Simon and Schuster "Golden Nature Guide" entitled INSECTS. Because of its colors this book is especially helpful. It also states clearly which insects are the bad ones, and which are beneficial and why. Don't be alarmed by the size of the illustrations. To anyone who may be a bit squeamish about insects at best, it may be terrifying to be confronted by a picture of a Lady Bug as big as a housewren! But there are figures in fine print which reassure us that the Lady Bug is 3/10 of an inch long in real life.

A more technical work is Lutz's FIELD BOOK OF INSECTS which makes it possible to identify insects not included in the more popular book.

The Year Book of the Department of Agriculture for the year 1952 is entitled INSECTS. This has excellent colored illustrations giving good detail of the various stages and metamorphoses of each species. This book recommends some rather drastic methods of control....

In our book the size of an insect is indicated by a straight line giving the actual size of the adult. If there is no line, the insect has been drawn life size.

Insect control requires identification. In order to identify, it is necessary to know about the various stages of insect development. Moths and butterflies, beetles, flies and insects of the bee, wasp and ant variety have four stages of growth: egg, larva, pupa and adult. It is most important to recognize at least the larval and adult stages because one is tempted to hand pick and destroy insects one can catch in the garden. One which is unutterably ugly to look at is the larva of the Lady Bug. We would be most ungrateful if we harmed these larvae in any way, because they devour aphids. Conversely the larvae of the Colorado Potato Bug are the ones that make skeletons of the potato plants. Oftentimes the insect which is destructive to garden plants in the caterpillar or larva stage, is most helpful, (for instance, with pollination) when it is in the butterfly stage. It should be noted also that although birds may not eat the adult of a species —a tough beetle, for instance—they avidly go for the same insect in its larval or grub stage. Therefore it is important to make a careful study in a good insect book to become well acquainted with all stages of metamorphosis.

Some species of insects do not go through a metamorphosis. These develop gradually in the same shape, and only their size changes. Grasshoppers, earwigs, members of the true "bug" family and some others, have what is known as "gradual development": that is, they start as miniatures of their parents and grow slowly to adult size without going through any metamorphic changes.

Some insect books carry a rather cryptogrammic method of indicating male and female adults of insect species. The male is indicated by the planetary symbol for Mars ♂ and the female by the symbol for the planet Venus ♀. The same system will be found in this book.

Many adults are already too settled in their dislike of insects to want to know them any more intimately. They are likely to panic when any insect, good or bad is mentioned. Most children, on the other hand, until they are conditioned by adults, are eager to appreciate the beauty of the blue-green beetle or the smoothness of the June Bug's wings, or the velvet wings of the butterfly. If these children can preserve their innate feeling of reverence and of kinship with the fantastic and exquisite and whimsical creatures of the insect world, they may grow up to be the naturalists who can help us learn more about biological controls of insects, and to replace some of the chemicals now under suspicion.

One faces today a multitude of choices among the insecticides. New ones appear each season and it takes a highly specialized student to really understand the fine print on the labels.

It depends on one's attitude, whether to use these inventions of chemistry in one's home and garden or to choose some of the substances and processes already well tested by time, and well known to be relatively harmless. If one understands the necessity for reverence toward all living entities, it follows naturally that one would prefer to use the latter.

Not all insect control is accomplished by substances, either chemical or biological. Many garden insects are controlled simply by the creation of maximum biological efficiency in the plants themselves. Healthy plants may have a few insects but they are not completely devastated by them. An example we see in our garden year after year: some flea beetles early in the season eat small holes in turnip greens and tomato leaves. But their season is short, and long before the turnips are formed in the ground or the fruit is developed on the tomato plant, the flea beetles have vanished. If the plants were not strong enough to resist a few small holes in their leaves, the damage would be important. But with well-nurtured plants the damage is not important enough to notice. The living vitality of the plant gives it strength enough to outlive the beetle damage: a process of vitality shortly overcomes any weakening caused by these insects in particular. This is only one example of insect control by a *process* rather than by a *substance*. We could cite numerous examples which we see in the garden year after year, but, of course, our garden is small and these examples are easy to overlook in a larger one.

Another kind of insect control, also admirably suited to the small back yard garden, is the use of predacious insects to prey upon the "bad." A friend once jokingly said to us, "When I think of your garden, I think of it as an example of organized mayhem." The insects which specialize in eating plants are eaten in turn by insect-eating insects: predators like the preying mantis or the larval stage of Lady-Bug and lace wing fly which eat aphis, or the parasitic braconid wasps that make their home in or on the bodies of other live insects, finally destroying them.

Without insect life on this earth, there probably would be no vegetation, just as without the bacteria in the human body, we could not live. Not all bacteria are injurious: there are, for instance, numerous bacteria which are absolutely necessary to processes of digestion. It is the same with insects. In the USDA Yearbook for 1952, INSECTS, they estimate there may be as many as 1,500,000 kinds of insects inhabiting our world. Of these, the injurious species in the United States may number only 6,500 to possibly 10,000. But such figures are unsatisfactory because no one really knows how many there are. Furthermore, some insects may be harmful in one stage but beneficial in another part of their life cycle, as explained above.

Students of the Bio-Dynamic Method since its very beginning have tried to accept and study insects whenever they have appeared, to learn from them further facts about the interrelationships which exist in living nature. It is now widely known that the interactions of living creatures, in *process,* are not to be learned solely from laboratory analysis or from the study of those creatures isolated from the rest of their environment, in *substance,* only. The old medical quotation "*In vito, non in vitro*" referred to this principle: in real life, not in a glass test tube, to translate freely.

In our own backyard gardens for the past thirty years we have, as amateurs, been making a careful study of all forms of life, from the soil bacteria in the root zone to the insects in the lower plant zones to the birds in the higher atmosphere, all of which contribute their part to the great living whole that is focused for human use in the growth and fruition of plants. The annihilation of any species of insect damages the totality of the interrelationships. Even mosquitoes feed fish and bats and darning needles!

To tell the truth, there have been some emergencies when we were tempted to use more violent controls. The section on Blister Beetles tells of a harrowing experience with a horde of devouring insects. For three years in Missouri we experienced drought on a 400-acre farm, and then the grasshoppers moved in and ate everything in their path (except the horehound plants). This became a serious economic problem. The farm superintendent could not use poison sprays because they might harm the cattle who would be eating the forage. His decision was to make silage of everything—"Grasshoppers and all!"

Using innumerable herbs and other plants, we made a study of grasshoppers' tastes, and tried to find substances which would repel them. We tried bitter herbs, castor oil, soap, urine, clay—anything we could think of that might work.

But there are always so many good insects that should be left alone that some of us have tried to learn specific treatments to take care of the bad ones without creating a condition that would be lethal to the good. Even pyrethrum sprays, which are relatively harmless to animals and men, still are so potent that they would destroy the good along with the bad. The same is true of every other poisonous substance. It seems wiser to study the situation and to learn as much as possible about existing conditions indicated by the presence of insect infestation. Oftentimes a "spell of weather" will cause certain insects to multiply temporarily, but when the weather changes they are dissatisfied and will leave. Sometimes a small infestation will be wiped out suddenly by the parent birds whose eggs have just hatched, and they suddenly have a nest-full of hungry babies to feed.

It has also been proved that creating maximum growth conditions in the fertilizing of plants or trees has relieved trouble from insect pests. Please notice the word *relieved.* The insects were not completely exterminated. Many times a healthier condition in the garden results in control of insect infestations, for instance, changing from a monoculture which encourages widespread insect ravages to the mixed culture recommended for the natural garden as the usual practice. It is well known that the mixed culture of plants in a small garden helps keep the insects under control, and this can even be done in large scale farming with the right crops.

For example alternating rows of potatoes and green beans helps keep both the Colorado potato beetle and the Mexican bean beetle under control. The book *COMPANION PLANTS* goes into this subject in detail and is recommended for anyone who wants to study this subject thoroughly. There are many beneficial combinations of plants still to be discovered and employed. Some are to be found in the old books on agriculture. For instance, in 1824 Loudon wrote:

"If in a patch of ground where cabbages are to be planted some hemp-seed be sown all round the edge, in the spring, the strong smell which that plant gives in vapor, will prevent the butterfly from infesting the cabbages. The Russian peasantry, in those provinces where hemp is cultivated, have their cabbages within those fields, by which they are free from caterpillars."

Some of us prefer to proceed with caution in the use of strong modern measures to kill insects. These chemicals in the long run may not be as helpful as they seem to be at first. It is generally known now what has happened with the housefly. The early use of DDT disposed of all but a few flies. The few that escaped bred more generations which were resistant to DDT. Now there are so many which can resist it that scientists have to invent new and stronger sprays. As Miss Carson pointed out in SILENT SPRING, this is the signal for the inventors to bring out stronger chemicals, and so on ad infinitum. Ultimately this may not be very helpful to the human race!

In the older gardening books there is a sense of calm and comfort. These were written at a time when people accepted insects as a part of the total landscape, as we, with the help of the Audubon Society accept birds. Our forefathers showed great wonder and interest in the insects, and wrote to gardening magazines, telling the discoveries they had made in their observations and experiments. In these letters one finds an attitude of curiosity, wonder and polite amazement. Our American attitude of horror toward a single housefly has been developed within the recent past. We have become so germ conscious and so obsessed with a passion for cleanliness that we are in danger of falling over backwards and ruining our chances for a wholesome agriculture, by destroying vast quantities of creatures, the good along with the bad.

From the perspective of these older writings, it is reassuring to read that in those days, in the section headed "Insects" they listed only aphis, caterpillars, slugs and snails (which latter two are not insects at all!) After reading present-day books on insect control, one is overwhelmed with their numbers and their omnipresence, and is saddened by the thought that man must try to exterminate them instead of learning how to keep them under control but still to enjoy the good they can bring to agriculture. This will take great patience and deep appreciation, added to our already great scientific knowledge and inventiveness. One is reminded of Dr. Steiner's observation that man may in time gain so much purely intellectual knowledge that it will do him but little good until he learns how to temper that intellectuality with warmer feeling and heartfelt attitudes.

The following pages are a presentation of some old and tested methods of insect control which might be applied to present day insect problems on the small scale of the backyard gardener. Most of these home remedies have been used either in our own garden or in the gardens of our friends. There is so much available information on this subject that one is tempted to write a small book on each phase. However, this tries in a modest way to cover the most common insects, especially the ones we have met in person during the past thirty happy years of backyard gardening.

what to do when you discover insect damage in your garden

First Study the insect through a magnifying glass to determine what stage it is in. Is it a caterpillar (or worm) or a butterfly or beetle or perhaps the larva of a beetle (See illustration of Lady Bug Larva). Any good illustrated insect book will help you recognize and identify your "bug."

Second Study the insect with your own observation to learn its food and its eating habits. Generally an insect eating one kind of plant will ignore all other plants not of that family. For instance, cabbage worms will eat cabbage, broccoli, cauliflower (all in the cole family) but they will not eat rose leaves. Similarly, the carrot worm will eat carrots, parsley, dill (any of the unbelliferous plants) but it will not be found on plants of other families.

Third Refer to the *vegetable, flower* and *fruit* lists for clues or the commonly found pests. When damage is discovered in your garden, check to see if it was done by one of the following, which commonly attack several different kinds of plants:

- aphis
- asiatic garden beetles
- earwigs
- grasshoppers
- red spider mites
- slugs
- sow bugs

Fourth Now that you know what the insect is, its name, habits and favorite plant food, look it up in this book in the Alphabetical List of Insects. Locate it in the following section to learn what non-toxic ways there are to keep it under control.

The eating habits of the insect give the clue to its identity and also to its control.

There are two main ways for an insect to feed: *Chewing* or *Sucking*.

chewing or biting insects

Examples: caterpillars, flea beetles, potato bugs, cankerworm on apples and roses, cutworms, grasshoppers, and a thousand others.

How they eat: These bite or chew the foliage or stem or the root.

CONTROLS

1. Repellents: aromatic or distasteful substances to keep them away
2. Handpicking and destruction by various methods
    a. Drop into a small jar with kerosene
    b. Collect and feed to a flock of hens
    c. Burn
    d. Drop into jar with water—See Hand Picking to Make Spray of Decomposed Insects
3. Predatory Insects: Preying Mantis, Lady Bug, Ichneumon Fly, Syrphid Fly, Ground Beetles, Lace Wing Fly
4. Wild Birds attracted to the garden by bird houses and bird-baths
5. Poisons applied to foliage to kill the insect that eats the leaf—only as a last resort.

sucking insects

Examples: aphis, thrips, nymphs of the squash bug, flies, scale insects.

How they eat: They suck the juices out of the plant, usually near the tender new growth at the tips of the branches. They are usually too tiny and too numerous to hand pick. They are not at all disturbed by stomach poisons applied to foliage. It is therefore necessary to reach them through contact with their outer layers, or some other methods.

CONTROLLING THEM

1. Predatory insects
2. Wild birds

3. Wash plant stems with cold water from the hose in forceful stream. Aphis are usually too tender to climb up stems from the ground.
4. Brush off with soft-bristled brush, holding plant stem supported along your own forearm and hand.
5. Asphyxiate tender bodies with any substance which will coat it and prevent its respiration through body spiracles or breathing holes.
   a. thin glue solution*
   b. waterglass solution*
   c. dilute clay solution*
   d. mustard flour*
   e. cold or cool water and soap suds—not detergents which would damage the plant as well as the insects
   f. stinging nettle brew* for black aphis
   g. quassia brew*
6. Simulate distasteful weather conditions. Find out what conditions make the particular insect thrive, and contrive some way to create the opposite kind of "weather."
   a. wind—shaking or blowing the insect with a cold stream of air
   b. rain—water from the hose

a prime method of insect control

. . . is to sow seed of a crop at a time when that crop's pest is inactive. For instance, sow carrots early when it is too cold for the root maggot fly to be abroad. Or sow carrots too late, when the maggots have already reached the fly stage. Then it will be too late for the fly to lay eggs for that season.

Corn sown early in the season, using an early-maturing variety, which should be ripe before the corn borer gets started.

Plant beans so that their blossoming stage does not coincide with the flying period of the adult bean weevil. Otherwise the adult weevil will lay her eggs in the blossoms. The weevils will mature inside the green beans causing some damage to the plant. The worst damage however occurs when the stored dried beans contain weevils which eat great holes in them.

One needs familiarity with both the crop and insects' habits and rhythms to make this kind of control a success.

*see by name under *Recipes and Formulas*

alphabetical lists of vegetables, fruits, and herbs, their common pests and companion plants

Vegetable	*Insects*	*Companion Plants*
Artichoke, Globe	a few aphis	
Artichoke, Jerusalem	none	
Asparagus	12-spotted and blue-black asparagus beetles	
Beans, Broad (Fava)	black fly	
Beans, Green Bush & Wax	Mexican bean beetle, root aphis, bean weevil	summer savory, strawberries, potatoes, beets, leeks, celeriac, summer radish
Beans, Pole	Mexican bean beetle, bean weevil	
Beans, Bush Lima	Mexican bean beetle, black fly	
Beans, Pole Lima	same as bush lima	beets, leeks, celeriac, radish
Beets	leaf miner	kohlrabi, leeks, bush beans, soy beans
Broccoli	cabbage worm, aphis, other cabbage insects	onions, aromatic herbs
Brussel Sprouts	same as cabbage	same as cabbage

Vegetable	*Insects*	*Companion Plants*
Cabbage	cutworm, harlequin bug, slugs, cabbage worm, cabbage root maggot	aromatic herbs, sage, rosemary, thyme, marjoram, oregano, onions, chives, garlic
Cantaloupe	see muskmelon	
Carrots	root aphis, harlequin bug, carrot worm, carrot root fly, wire worm	chives, lettuce, peas
Cauliflower	same as cabbage	same as cabbage
Celeriac	celery hopper	
Celery	celery hopper, tarnished plant bug	
Chicory	few or none	sow in pea rows after pea harvest
Chinese Cabbage	same as cabbage	same as cabbage
Chives	none or few	
Collards	same as for cabbage	same as for cabbage
Corn salad	none or few	
Corn, Sweet	European corn borer, corn ear worm	
Cucumber	aphis, striped cucumber beetle, root maggot fly	bush beans, corn, dill, lettuce, radish
Dandelion	none or few	
Dill	carrot worm	
Eggplant	flea beetle, Colorado potato beetle cutworms	
Cress	none or few	
Endive	none or few	
Garlic	onion maggot, grey fly larva	roses
Horseradish	flea beetles	
Kale	same as for cabbage	same as for cabbage
Kohlrabi	same as for cabbage	same as for cabbage
Leek	onion maggot, grey fly larva	bush beans, beets, celeriac, carrots
Lettuce	slugs, cutworms, root aphis	red cabbage, Brussels sprouts, radish, cabbage, broccoli, shallots
Muskmelon	striped cucumber beetle, squash bug, squash vine borer	radish
Mustard Greens	same as for cabbage	same as for cabbage
New Zealand Spinach	few or none	

Vegetable	*Insects*	*Companion Plants*
Okra	green stink bug, cabbage loopers	
Onion	onion fly, onion maggot, thrips	all of cabbage family
Parsley	parsley (carrot) worm	asparagus, celery, leeks, peas, roses
Parsnip	carrot worm	tomatoes
Peanut	none or few	
Peas	pea aphid, red spider mite, bean weevil	carrots, potatoes
Peppers, Hot	cutworms	do not plant hot and sweet peppers together
Peppers, Sweet	cutworms	
Potato	old fashioned potato beetle, flea beetle, Colorado potato beetle, wireworm, potato stem borer	horseradish
Pumpkin	squash bug, squash vine borer	corn, radish
Radish	radish root maggot, flea beetle, harlequin bug	beans, pole & bush, kohlrabi
Rhubarb	few or none	
Rutabaga	same as for cabbage	same as for cabbage
Salsify	carrot worm	
Soybeans	few or none	
Spinach	leaf miner, flea beetle	strawberries
Squash, Summer	squash bug, squash vine borer	nasturtiums, marigolds, radish
Squash, Winter	squash bug, squash vine borer	nasturtiums, marigolds, radish
Sunflower	stem borers	
Sweet Potato	few or none	
Swiss Chard	leaf miner, grasshoppers	marigolds, nasturtiums
Tomato	cutworm, flea beetle, tomato horn worm	stinging nettle, parsley chives, asparagus. Bad companions are fennel and the cabbage family
Turnip	sow bugs, flea beetles, cabbage worm, harlequin bug	
Watermelon	cucumber beetles, same as for squash	nasturtiums, radish, marigolds

Insects Found in the Herb Garden

Calendula	cutworm, climbing cutworm
Dill	carrot worm
Marigold	greenhouse white fly
Nasturtium	aphis
Parsley	carrot worm
Peppermint	greenhouse white fly
Sage	hawk moth caterpillar
Spearmint	hawk moth caterpillar
Sweet Basil	Asiatic garden beetle

Insects Attacking Small Fruit

Blackberry	aphids, cutworms, Japanese beetle, cane borer, galls
Blueberry	blueberry maggot, Japanese beetle, galls
Currant	current aphid, currant worm, gooseberry caterpillar
Elderberry	none or few
Gooseberry	same as currant
Grapes	leaf hoppers, Japanese beetle, rose chafer, grape curculio, leaf tier, mealy bug
Raspberry	raspberry fruit worm, cane borer, rose chafer, white grub, red spider, grasshopper
Thimbleberry	cane borer
Strawberry	cane borer cutworm, crown borer, sawfly, strawberry weevil, curculio

Insects attacking Small Fruit Trees

Apples	coddling moth, curculio, tent caterpillar, cankerworm, apple maggot, European red mite
Apricot	same as for peach
Cherry	plum curculio, tent caterpillar
Cherry, Wild	tent caterpillar, leaf tier
Peach	plum curculio, peach borer, Oriental fruit moth
Pear	plum curculio, pear psylla, coddling moth, rose slug
Plum	plum curculio
Quince	coddling moth

Insects Attacking Common Shrubs

Hydrangea	woolly bear caterpillar
Laurel	European leaf roller
Lilac	European leaf roller, greenhouse white fly, leaf miner
Privet	leaf miner
Rose	European leaf roller

Insects Frequently Found in the Flower Garden

Alyssum	cabbage worm
Aster	root aphis, flea beetle, leaf tier, greenhouse white fly, earwig
Calendula	cutworm, climbing cutworm, Japanese beetle
Carnation	cabbage looper
Chrysanthemum	cabbage looper, flea beetle, gall
Columbine	columbine leaf miner
Cosmos	few or none
Dahlia	corn ear worm, burdock borer, earwigs
Geranium	cabbage looper, leaf tier, fall web worm, tussock moth caterpillar
Gladiolus	thrips
Hollyhock	burdock borer, slugs, iris borer
Larkspur	burdock borer, leaf miner
Marigold	Japanese beetle, earwigs
Mignonette	cabbage worm, cabbage looper
Nasturtium	leaf tier, aphis, greenhouse white fly
Nicotiana	Colorado potato beetle, tomato horn worm
Night-scented Stock	cabbage family insects, leaf miner
Pansy	leaf miner, cutworm
Peony	leaf roller
Phlox	corn ear worm
Petunia	flea beetle
Rose	leaf tier, rose chafer, rose leafhopper, burdock borer, fall webworm, white-marked tussock moth, tent caterpillar, corn earworm, rose slug, rose saw fly, rose flea beetle, June bugs, earwig, leaf gall
Snapdragons	stink bugs
Sweet Pea	pea aphid, corn ear worm, red spider mite
White Daisy	thrip, earwigs
Zinnia	tarnished plant bug, Japanese beetle

alphabetical list of insects.
their damages and their controls

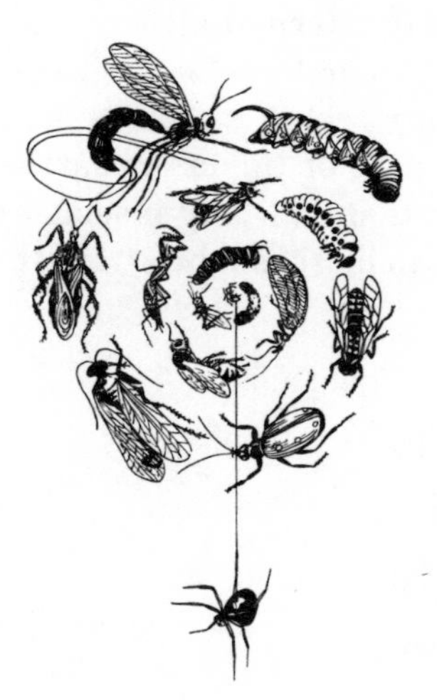

Note: Details on control appear in the *Recipes & Formulas* section. These are referenced here by an asterisk.

ant

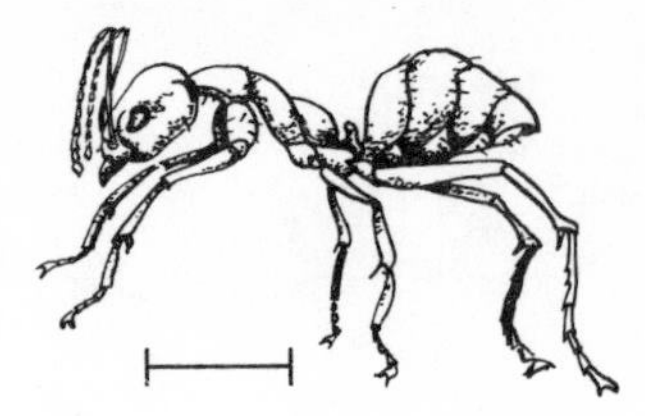

In the Bio-Dynamic garden with raised beds the ants have never done any damage, scurrying happily about their work in their burrows in the paths. It is possible to prevent ants from getting into fruit trees by smearing a sticky material around the trunk. Ants can be kept out of bee hives by sprinkling red pepper in the upper cover where they try to enter. The bees themselves prevent their coming in through the hive entrance. If ants are troublesome in the house, there are several harmless ways to repel them: Pennyroyal or tansy leaves scattered on the kitchen shelves, or egg shells broken up and scattered wherever the ants are found. Red pepper and borax repel them. Some use drugstore Creolin. Dilute according to directions and pour around house on all foundations.

Red Ants in the house may be driven away by scattered Sweet Fern (which is not really a fern at all, but a small shrub identified as *Comptonia Asplenifolia*) or by a small bag of sulphur in drawers and cupboards. An old recipe suggests a quart of boiling water poured over one half pint of tar in an earthen vessel and left in a closet to drive away red ants. Many insects are repelled by the odor of tar, and heat serves to increase the strength of that odor.

aphis or green fly, white fly, black fly, plant louse

The aphid or aphis is a most familiar small pest, both in the garden and on house plants. It is so small that a magnifying glass is almost a necessity to detect its presence. When newly-budding terminal leaves and stems begin to look pale and spindly, look first of all for aphis (or aphides—both spellings are accepted.) Sometimes they are called green, black, or white fly, or plant lice. They come in green to match plant stems, or in red or in black. They are sucking insects and they affect nearly all plants at one time or another. It should be noted that they attack house plants late in the winter season when the plants' resistance is low; they often appear on garden plants when the weather begins to settle in for the heat of the summer.

They pass through various stages from nymph to adult, both with wings and without wings. When the aphis on one plant become over-crowded, they quite suddenly develop wings and fly to another host plant of the same plant family. They mature in about 12 days and there may be an infinite number of generations each season. In fact, if it were not for the insects and birds who devour these aphis in astronomical numbers, much vegetation would be damaged or destroyed by them every year.

However, nature in her marvelous way has arranged that the fast-multiplying aphis are prey for other insect eating insects: the Lady Bug beetle with her weird looking larva is always on the lookout for aphis and these larvae help to keep the aphis under control. Whenever a bright scarlet Lady Bug with black spots is visible, it is a sign there are aphis on that plant. We include here a sketch of the Lady Bug larva because it is such a repulsive looking little creature that many are unwittingly killed.

The Lady Bug larva* is a *good* bug. See Lady Bug.

Root Aphis may appear on carrot roots, lettuce, parsley, even green beans and peanuts. Use a strong brew of onion* and pour around the plants when the soil is moist. If a stronger measure must be taken, try a solution of rotenone and pour it directly over the plants so as to reach the roots without soaking adjacent soils. If a still stronger measure is indicated, loosen the soil around affected roots and apply nicotine sulfate solution diluted: one teaspoonful to one gallon of soapy water.

When broad beans are attacked by black flies (as they always are) cut off the stems below the infestation and burn. The plants will outgrow the amputation. See also White Fly and Woolly Aphis.

aphis lion or lace wing fly

Another insect preys upon aphis with such voracity that it is even named the Aphis-lion. This is a pale greenish fly-like insect often seen flying about the vegetable garden in summer. The larva is slender with very long curved jaws with which it seizes and devours the tender body of the aphis. The aphis-lion lays her eggs in an unusual way on the leaves of a plant up on the ends of long hairs to prevent them from being eaten. This is another insect which deserves our admiration and protection in the garden because it works steadily in our interests. This is more often called the lace wing fly.

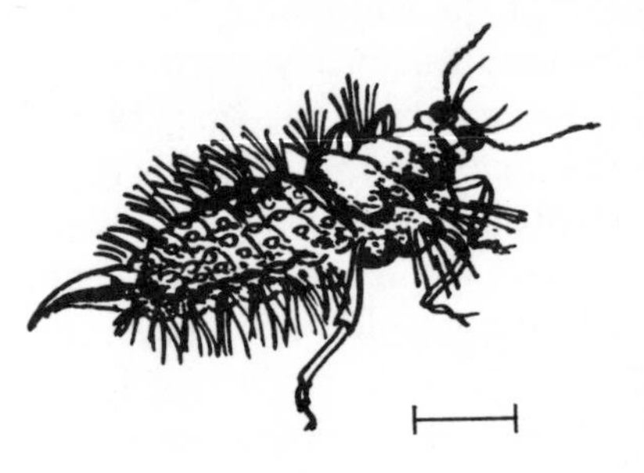

APHIS LION larva

Still another insect which relishes juicy plant lice or aphis is the larva of the sweat fly or hover fly. This larva is known to devour the aphis one by one.

In searching for harmless ways to control the aphis which visit all gardens in greater or lesser numbers, usually attacking plants past their prime, we have tried to find practices which will not do too much harm to the Lady Bugs, lace-wing flies and the hover flies. A few recommendations follow, some of them aimed specifically at the Bio-Dynamic Garden because they must, in order to be effective, be used in conjunction with other Bio-Dynamic substances or processes.*

1. Stinging Nettle Brew* and Equisetum Tea*
2. Support plant stem on your hand: carefully brush off aphis with soft bristled brush. Aphis usually appear at the very top of the stem.
3. Wash aphis off stems with a fast stream of cold water. Repeat often.
4. Dip house plants in warm water, not hotter than 140°F.
5. Cut off infested tops and burn them with aphis on them.
6. For aphis on apple trees, scrub bark with a stiff brush dipped in hot salt water.
7. Wash plants with cool water and soap suds made with a

mild soap, or soft soap. Do not use detergent suds as some may kill the plant too. Try fermented Equisetum tea* with soft soap added.

8. Spray aphis on a plant with a dilute solution of clay. The wet clay will kill the soft-bodied aphis.
9. Spray with a very thin mixture of glue which clings to the aphis' body and smothers it.
10. For very desperate situations aphis might be controlled by commercial nicotine preparations or by kerosene emulsion* or by tobacco smoke fumigation.*
11. We have used tobacco dust to good effect to get rid of aphis.
12. Quassia Solution* if made right, will prove helpful.

One more word about aphis. When we find them in the garden we do not hurry to destroy them. We prefer to let them work for a little while to see what they will do and to see if some natural predators will take care of them and save us the trouble. For instance, the scientists who study birds tell us that the House Sparrow is especially partial to aphis and eats more than any other bird.

One summer we found some green aphis on one branch of an apple tree. We kept an eye on them to see that they did not spread unduly. Before long we could see that the birds had found them too. Within a week there was not an aphis left on the apple tree. It was no trouble at all for the birds to find and dispatch every single aphis and we were glad that we had not tried to destroy them ourselves.

The appearance of aphis seems to mark a place in the plant's cycle of growth. Often the aphis appear when the plant is weakened by its flowering period: like a human being past the prime of life who is susceptible to many small ailments. Or the plant may be weakened by an adverse "spell of weather" like drought which weakens the plant's nutrition. It is possible to brace up the plant in the latter case, by a special feeding of compost* or compost water* or Nettle Brew*. If we simply kill off the aphis, we remove one of the indicators which point to the inner condition of the plant. Admittedly, it takes much courage and patience and a few years of experience before one is willing to stand quietly by and watch an infestation without a good deal of misgiving. Sometimes it is a hard test of faith to leave the serious job of controlling aphis to the birds. But we recommend this restraint and patient observation as a way to learn more intimately the processes and secrets of the growth of plants.

army worm

Sometimes one insect species will get ahead of its insect parasites and other fungus or bacterial ills which might keep it within bounds. It may then multiply to such an extent that its larvae travel in hordes eating everything edible in their path. The Army Worm is so named because it is known to travel like an army, en masse, devouring everything in a wide swath. The larvae, which most of us would call caterpillars or worms look like the illustration, about two inches long, grayish-black with three longitudinal yellow stripes on the back. They feed at night. During the day they hide in the grass roots in the soil. There are two or three broods a year. They make their cocoons in the topsoil and adult moths emerge in about two weeks. The adult females may lay as many as seven hundred eggs apiece.

In case of an infestation of army worms, it is possible to stop them by digging a ditch around the field. The side next the field should be plowed so it is perpendicular so the worms cannot crawl out. They will tumble down into the ditch and may be crushed with rollers or buried deep with earth or burned by filling the ditch with straw, pouring on kerosene and then igniting it. These are strong measures, but army worms are an emergency and a serious menace when they suddenly arrive. Fortunately for long periods they do not appear in any locality.

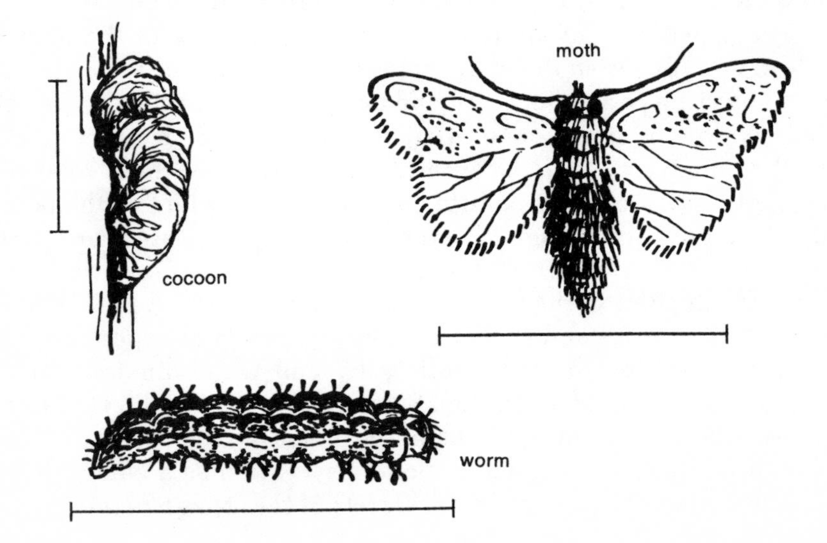

asiatic garden beetle

In recent years we have been seeing more often a small brown beetle a little less than one half inch long, which looks harmless enough in itself. One year we had four half grown basil plants which lost a leaf or two every night. Finally before the plants vanished completely, we kept watch off and on, day and night. Early evening brought the solution to the mystery. The Asiatic Garden Beetles were rapidly devouring the crisp green leaves.

The adult beetles may be attracted by an electronic trap*. The grub stage, if you find them in the soil, can be treated with a contact insecticide such as derris* or rotenone*.

asparagus beetle

The Asparagus Beetle is a small red or green insect with six or twelve tiny spots on its back. It eats the stalks of the asparagus. It can do considerable damage and of course, poisonous sprays should not be used on an edible crop. One very simple way to overcome the asparagus beetle is to fence in the asparagus bed and let hens run in the area. If some grass sod is also enclosed, the hens will not eat the asparagus plants. But their sharp eyes will detect every single beetle and they will waste no time in cleaning up the entire beetle population. If the beetles are controlled during late summer and fall of one season, it means that the crop will be free of damage the following spring. There is also a blue-black asparagus beetle with three white spots and an orange margin on each wing.

It has been stated in some recent "science reporting" that planting tomatoes in the asparagus patch will repel the asparagus beetle, but during many years of actual practice we have never found this to work. The beetles flourished in spite of the tomato plants, but the hens and the wild birds kept them under control.

In fall, when the beetles were most active, the wild birds especially found the asparagus patch much to their liking, with sparrows and goldfinches actively foraging for beetles all day long.

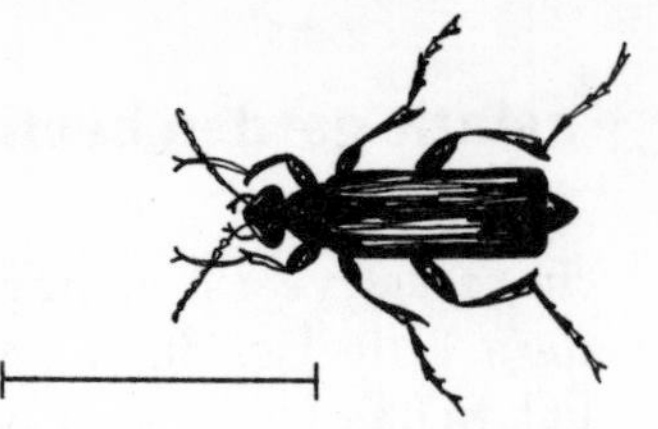

blister beetles

One Sunday afternoon our neighbor telephoned: "There's a most dreadful commotion going on in our garden—a host of gray and black beetles which are coming in a cloud and devouring everything in sight." We grabbed the bug book and ran. She was right. They were marching straight ahead, thousands of them, eating everything they touched. We found them in the book: Blister Beetles, so named because one caught in one's clothing or under a watch bracelet will raise a blister. (We tried that too!) They are related to Spanish Fly which used to be used in medicine for blistering.

Sometimes a horde of live invaders can be stopped by a stronger horde. Our first thought was to organize a regiment of birds: the Hens! We lured the flock with grain to the spot where the beetles were swarming down the Swiss chard row and eating it down to the roots. One hen ate one beetle and that was all. How she reacted inside, or how she communicated to the rest of the flock we never knew. But it was evident that hens will not eat blister beetles.

One source suggested that such an invasion can be driven. That was what we did. We enlisted as many people as we could find, armed with sticks and branches and all making a loud noise and sweeping motions. We literally drove that swarm of beetles right through the garden and out into the fields on the other side. And they never came back.

borers

The Borer is an unsporting kind of insect. He hatches out inside the stem and eats and grows and grows and eats and no one knows he is there. In June or July suddenly the top of the plant withers. Various kinds of borers attack corn, larkspur, hollyhock, dahlia, rose, iris, potatoes, columbine and sunflowers.

The burdock borer is commonly found in burdock stems. It is a smooth, pale-brownish caterpillar with a whitish stripe down the middle of its back and on its side a band about twice as wide of the same color. It attacks the plants named above.

The iris borer when mature is about 1½ inches long; flesh colored, and it looks like a cutworm—slightly pinkish along its back.

Whatever the plant, whatever borer is responsible, cut off the injured stems and burn them to destroy the borers. In the fall burn all old stalks to destroy the eggs. See Squash Vine Borer.

braconid wasp

Our first acquaintance with the phenomenon of the Braconid Wasp occurred during an inquisitive childhood when a friendly adult took considerable care in setting up demonstrations of living insects so that we could observe what would happen. We had some milkweed plants with caterpillars eating happily on the leaves. The caterpillars in all probability belonged to the sphinx-moth family, although to us children the exact identification of the species was unimportant. Some of the caterpillars, after they had eaten their fill, changed right before our eyes into the characteristic pupal form. But one caterpillar suddenly blossomed out with tiny white specks which resembled eggs in shape. We watched them increase in number until there were thirty or forty down the caterpillar's back.

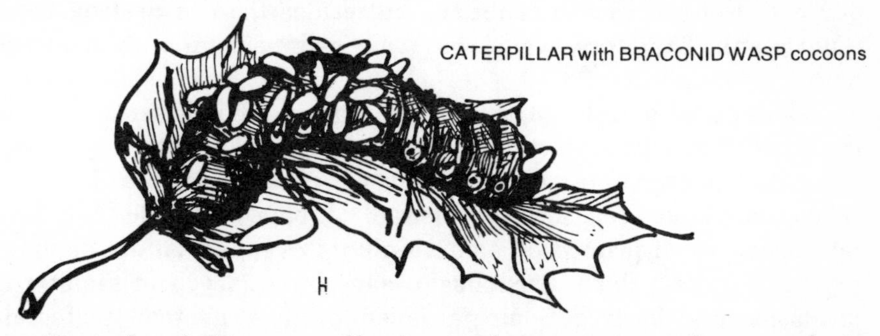

CATERPILLAR with BRACONID WASP cocoons

We isolated this caterpillar in another container—fortunately it was a large goldfish globe with a cover. The caterpillar by this time was obviously not feeling at all well. Within a day or two we observed that each white "egg" had been a tiny cocoon. Each one had a small "cover" which opened at the top of the cocoon. In no time at all the globe was filled with infinitesimal flying wasps. Later we learned what is now common knowledge, that the adult female wasp lays her eggs in the body of the caterpillar. As the wasp larvae hatch and begin to eat, the caterpillar is devoured and finally is killed by these parasites.

The rule is: never destroy a caterpillar which bears these visible wasp cocoons because the braconid wasps are a valuable help in the garden and should be encouraged and allowed to multiply. Any kind of poison spray will, of course, bring their good work to an untimely end.

cabbage butterfly

Fortunately there are many common garden insects with which one has unspectacular encounters. The Cabbage Butterfly is the only butterfly that seriously injures any of our familiar garden crops. The female looks dainty and attractive as she flits innocently from leaf to leaf around the cabbage and broccoli and cauliflowers. *But*—within a few days tiny green larvae start to hatch from the egg clusters she was delicately laying each time she dipped to touch a leaf. In almost no time, it seems, the little green larvae have made lace skeletons of the cabbage leaves unless they are kept under control.

It is quite practical to crush the egg clusters with one's fingers wherever one sees them. Since cabbage is an edible leaf, it is unwise to spray with anything poisonous. It also seems more sporting to use our wits to overcome the cabbage butterflies than to destroy them point blank. Therefore, we have tried repelling them with aromatic herbs.

Herbs grown in the garden at the ends of the beds serve to repel the butterflies who seem to dislike aromatic herb odors. Sage, rosemary and peppermint are especially effective repellents. In years when there have been large numbers of cabbage butterflies, we have laid branches of pruned off tomato plants over individual cabbage plants to protect them. Branches of sage, rosemary, and the strong mints also may be used. Some gardeners prefer to destroy the female butterflies. One gardener of our acquaintance even goes after them

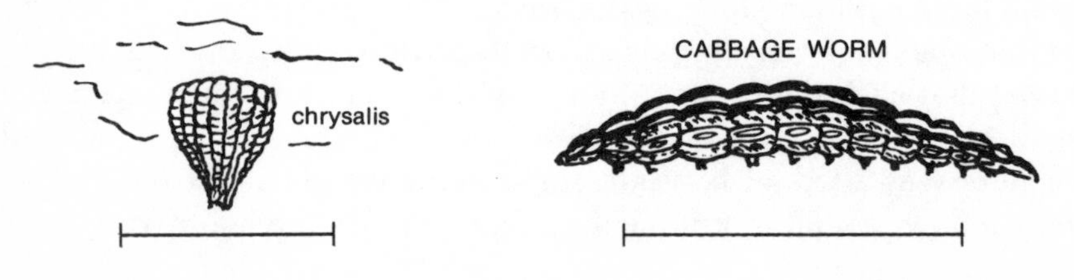

with a tennis racket! Another recommends powdered tansy* sprinkled on the cabbage heads.

When it comes to controlling the small green larvae, it takes vigilance and good eyesight. When the larva first hatches from its egg, it is very small. But it quickly grows and grows and grows, until it has reached full size. All the time it is a beautiful velvety jade green which exactly matches the green of the cabbage leaves.

One way to locate the worm is to look at the lower leaves for the droppings which have fallen as the worm chewed its way along the upper leaves. It is not difficult to hand pick the green worms and then either crush them or feed them to the birds.

One fine day the adult worm will stop eating and will transform itself into an exquisite, jewel-like chrysalis. Careful examination of the chrysalis will reveal characteristics of both the worm and the butterfly. The lower end echoes in form the segments of the worm, and also the undulating curves of the butterfly's body. The chrysalis hangs by a silken thread until the metamorphosis within is completed. Then the shell splits and the graceful butterfly emerges, ready to start the cycle of egg-laying all over again. There may be three broods in one year!

The cabbage worm devours all members of the cabbage family, but especially likes broccoli and cauliflower. It may be controlled by handpicking, by sprinkling with a weak salt solution* in water (the salt draws the vital fluids out through the delicate skin) or by sprinkling in the early morning with rye flour or with powdered lime while the worm is still cold and immobile. The flour or lime will dehydrate the insect and cause its collapse.

cabbage looper

The Cabbage Looper is found feeding on the underside of leaves of all the cabbage family. He is identified by the characteristic way he humps along like a measuring worm. The adult emerges from the pupa in spring as a brown moth. The Cabbage Looper may be controlled by the new organic insecticide called Dipel*.

Research has been done on the biological control, Polyhedrosis, which promises to destroy innumerable loopers each year. For more information on this new control see *Handbook on Biological Control of Plant Pests,* Brooklyn Botanic Garden Record, Plants and Gardens, Vol. 16, No. 3.

calosoma beetle

One of the most attractive families in the insect world is the group called Ground Beetles and their close relatives the Tiger Beetles. These are long-legged, hard shelled beetles which lope along over the garden soil as though on urgent errands. A famous member of this family, whose portrait is featured in every insect book is Calosoma Scrutator whose name means the "Searcher." He is so fond of caterpillars that he will even go up into a tree in search of them. The larva also feed on caterpillars. In fact shortly after the turn of the century this insect was imported from Europe to the eastern seaboard of the United States purposely to combat the Brown-tail Moth, and there was a fine of $5 for anyone who killed a Calosoma beetle. In 1962 we were glad to hear a government entymologist say that the Brown-tail Moth does not pose any great threat in America any longer.

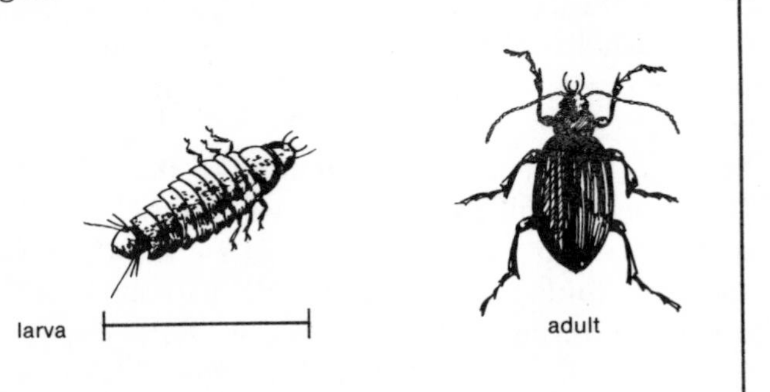

Calosoma has other features which make him well known to those who like to observe insects closely. His wing covers are a metallic greenblue, slightly ridged, with a brilliant red border. His six legs seem unusually long, but they provide his speed. He can be identified by the eleven fine joints in his very handsome antennae. For protection from those who would molest him, he gives off a very acrid odor which is so characteristic that he may be located in the garden by his scent alone. The larva has twelve segments, voracious jaws, just right for snatching and devouring caterpillars. He also has a pair of bristly hairs on his tail end, possibly to make him look ferocious and inedible from that angle also.

There are many many other busy members of this hardworking family and all are helpful in keeping harmful insects under control. A small one even eats potato beetles. These are among our most distinguished and most active predatory insects and they should never be harmed.

Even the Egyptians recognized and valued this insect family so highly that they carved scarab beetles in precious stone and used them as amulets and in the decoration of their most sacred buildings. The present family name of these beetles, Carabidae, preserves the historical connection with the knowledge and appreciation of the ancients.

The Calosoma beetle is inordinately fond of Tent Caterpillars and should be encouraged to eat his fill of them every season!

canker worm

The Cankerworms spin long silken threads and come twirling down from the apple trees at the exact time when you want to go out to enjoy the blossoms! At other times the larva, which is a large tender green worm, will straighten itself out and lie motionless, blending with its background, intending to make predatory insects and even human beings think it is only a small twig.

Eggs are laid in the tree tops by a wingless female which must climb up the trunk on her six slender legs. If she has to pass over a band of sticky material, her feet will be trapped and she will be prevented from laying eggs. See Bands on Fruit Trees.*

One choice bit of information about the Cankerworms and their kin is that the whole family is called Geometridae or "earth measurers." Who has not watched the typical progress of a measuring worm or "looper" as it is also called, as it pushes its hind three feet up to its front three feet, looping its entire body in the process?

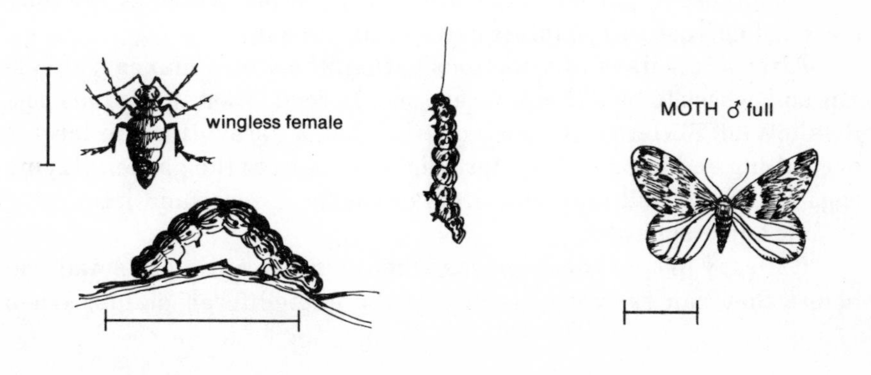

For control of the worm stage, see directions under Hand Picking and Letting the Insects Decompose.*

There is a spring cankerworm and a fall cankerworm, two different insects, each of which raises only one generation each year. Both have many natural enemies: tachinid flies, ground beetles, potter wasps (which stuff cells with cankerworms to feed their young), and predatory insects that suck the juices out of the delicate cankerworms. There are over 40 different kinds of birds that relish them. Sometimes even the early frosts finish them off prematurely.

The male adult of the cankerworm is one of those fluttery little moths attracted by the dozens to any light at night. A lighted insect trap* or moth bait* with a light above it will help control the cankerworm moth.

Another harmless method of control is to cultivate the soil under apple trees very thoroughly, because around the first of June the larvae spin down from the trees and enter the ground about one inch. There they are transformed into greenish-brown pupae. If the soil is then cultivated, the birds will come joyfully to eat the pupae, which will help control the larvae for the following spring season.

carrot worm, parsley worm or **black swallowtail butterfly**

The backyard gardener finds this dark worm most disturbing when he looks early in the morning at carrot or parsley plants and finds them thoroughly chewed. He will be still further disenchanted if he touches the velvety worm and it shoots out a pair of orange colored "horns" which at the same time emit a strong musky odor. Actually these apparent horns are a scent organ which is the only means of defense this helpless caterpillar can use.

After a few days of voracious eating, the worm makes a chrysalis and ceases to be a threat to anyone. In fact, when it emerges as a Swallowtail Butterfly, it is so beautiful that we are willing to forgive everything and to enjoy its charm as it floats over the garden (laying eggs for the second crop of worms to eat the carrots and parsley of the late summer garden!)

It is easy to spot these worms when they are still very small and before they can eat very heartily of the umbelliferae plants. When

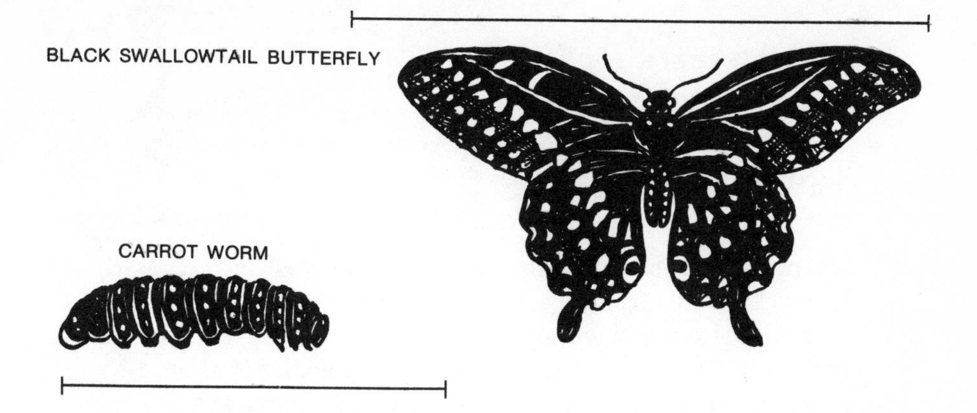

they are tiny, they look almost jet black. As they increase in size, the green stripes with yellow dots on the black bands also become apparent, and it becomes one of our best dressed caterpillars. Hand-picking early in the morning is still the easy way to keep this heavy feeder from doing very much harm in the home garden.

Carrot root maggot and wireworms attack carrots in the ground. To repel carrot root maggot try a creosote-soaked rope* along the carrot row. See also control for wireworm*.

caterpillars

There are many different kinds of Caterpillars because this is the name applied to the larval stage of many moths. (Our English name "caterpillar" comes from two Latin words: *catta* and *pilosa*, meaning "a hairy cat"!) One of the old books (Sutton) suggests that caterpillars can be reduced in numbers if one will look for patches of eggs and clusters of young caterpillars on the under sides of leaves and nip off those leaves and burn them. "This enemy cannot be raked in rank and file, but must be taken in detail as in guerilla warfare."

He also says "Caterpillars cannot often be treated in a wholesale way, because to reach them effectively is apt to endanger the plant. . . . Hence we are usually compelled to rely on hand-picking, and we are bound to observe that, tedious as this may be, a little patient perseverance will accomplish wonders."

What Sutton said nearly seventy-five years ago is still true today—a little patient perseverance will accomplish wonders.

coccus (see scale insect)

codling moth

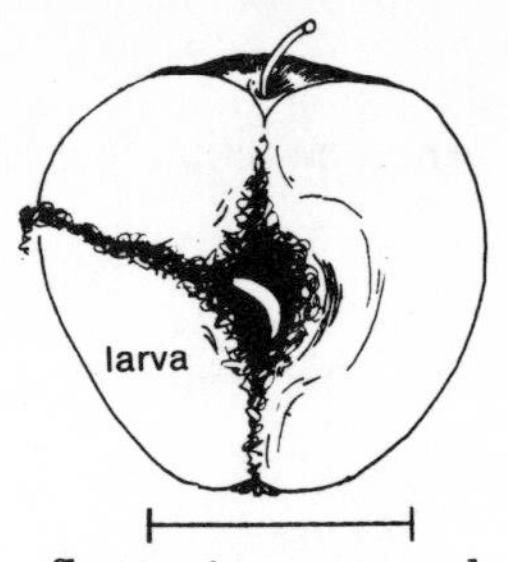

When this innocent-looking little moth comes fluttering around the electric light in spring, it seems a long time before the small pinkish worm appears when we bite into an apple in the fall. The Codling Moth is the mother of the apple worm. In passing it should be explained here that a codling is another name for a tiny young apple as it commences to take shape after blossoming.

When the young apple is forming, three or four weeks after the petals have fallen from the blossoms, if the moths can be prevented from laying eggs, the larvae will not enter the apples through the blossom end.

Better control of this insect depends on various kinds of work on fruit trees at different times of the apple growing season. Briefly: the moths may be attracted to various kinds of baited traps after which they can be destroyed. See Fruit Tree Baits*. They may be lured by lights into some sort of trap or electrocuting device. See Electronic Insect Traps*.

Another way to trap the species is to apply sticky bands in July to catch the larvae as they crawl down the tree after leaving the apples. See Bands,* Barrier. Remove the bands and destroy the worms about every 10 days. The larvae crawl downwards seeking to spin a cocoon in a protected place in the bark of the tree. A way to expose them is to scrub the bark with a rough brush to clean off all loose pieces of bark where the larvae might over-winter.

Some codling moths may fly out of the apple storage room in late winter and early spring. If the storage room is screened, the moths cannot escape into the apple trees to start laying eggs again.

colorado potato beetle

Most people are familiar with the brown and white striped beetle by this name, but many do not recognize the ugly reddish or pink 'slugs' as the larvae of this beetle. The best time to control the Colorado Potato Beetle is the day the first adult flies into the potato patch. Then it is easy to catch her and to crush her between two stones before she lays any eggs at all.

If you miss the first one, and the eggs are already laid, it is not hard to find the eggs on the underside of potato leaves and to crush them quickly.

If the eggs hatch out, the horrid little reddish larvae begin immediately to scatter and to make inroads on the potato leaves. They are also very fond of eggplant.

COLORADO POTATO BEETLE

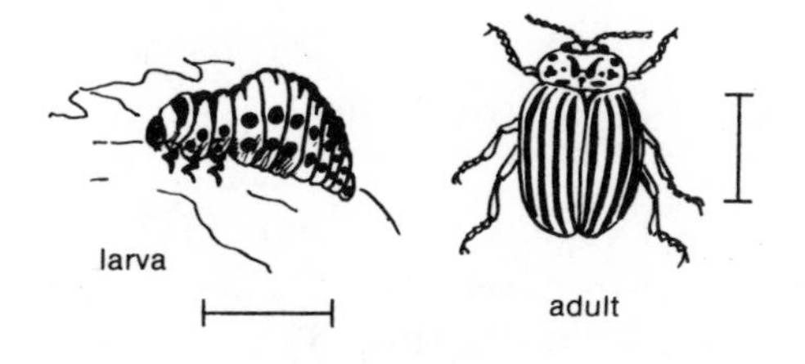

For many years in our gardens it has been routine practice to plant potatoes and bush beans in alternate rows. Something in the bush beans repels the Colorado Potato Beetle. As yet no one knows what it is, or why it works. But that it does work, we know. There is, however, some evidence to show that some garden conditions may be changing, as there have been more Colorado Potato Beetles during the past three or four years. Some say that petunias used to repel potato bugs. A flax plant set out at regular intervals in the potato patch will still repel the beetle.

Hand pick the larvae and drop them into a small can of kerosene regularly each day if the egg laying has a head start.

corn borer (european) and corn ear worm

Those dainty moths that flutter around the yard lights or dance before car headlights on a country road in a summer evening are heralds of gardening troubles ahead. The female moth lays her eggs in corn stalks and in many different weeds. The eggs hatch and the hungry larvae devour the plant substance around them, causing serious damage in the corn crop.

Just as the moths are drawn to car lights, they may be lured to an Electronic Insect Trap*, available from Burpee Seeds, P.O. Box

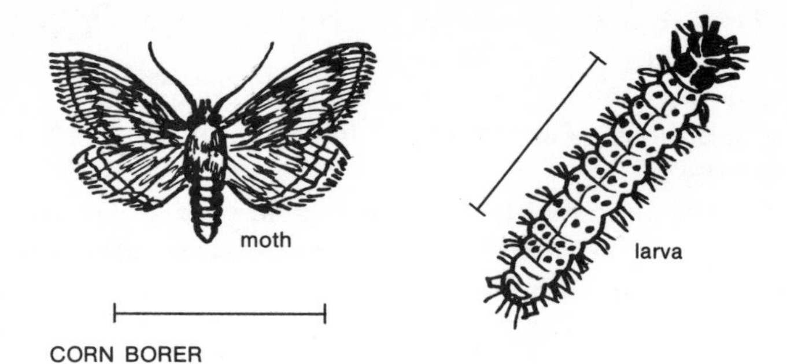

CORN BORER

6929, Philadelphia, Penna. 19131. They may also be attracted by Baits*.

Plowing corn stalks under at the end of the season helps expose and thus control the corn borer. Another method used to control insects like this borer is the Bacillus Thuringiensis* which is used commercially in large acreages as well as in backyard gardens.

The corn ear worm, which is also called tomato fruitworm, is another night-flying moth whose eggs are laid on cornsilk. The larvae hatch and eat their way down into the corn ear. Control is relatively simple. Take a long spouted oil can or a medicine dropper. Drop about a quarter-teaspoonful of mineral oil on the corn silk after it has wilted and turned brown but before it begins to dry, or add red pepper to oil to squirt into the cornsilk. One part mineral oil and one part pyrethrum extract* will be effective, using one quarter teaspoon per ear. The oil envelopes the larva and seals its breathing spiracles, thus suffocating it; or use Dipel*.

Odorless marigolds planted around and among the corn sometimes repel the corn ear worm.

The moth of the corn ear worm may be controlled by electronic insect traps* or by moth baits*. Corn ear worms and corn borers are also attacked by the Trichogramma wasps.

Pumpkins make a desirable companion crop with corn. The American Indians learned by their own observation that this combination resulted in healthy plants of both species.

CORN EARWORM

cucumber beetle, striped

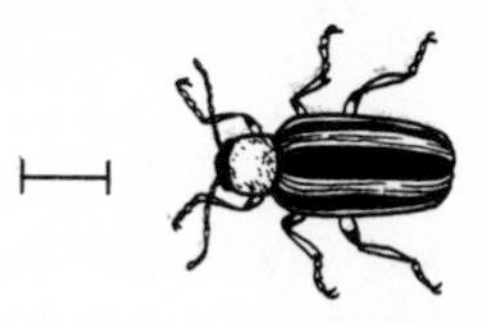

The small yellow and black striped Cucumber Beetle is not really as large and formadible as this illustration suggests. It is actually the length of the line drawn beside the magnified illustration.

This small but lively beetle is developed from white larvae which live among the roots of corn or plants of the melon-squash-cucumber family, and it does a great deal of damage.

Early in the season the beetle may be controlled by sowing radish seed in the same hill with cucumber or melon seed. There is something about radish that repels the little yellow and black striped beetles which come after the tender cucurbit seedlings.

If the adult beetles are repelled from the home garden, there will be no eggs or larvae to damage the roots. It is that simple!

curculio—plum

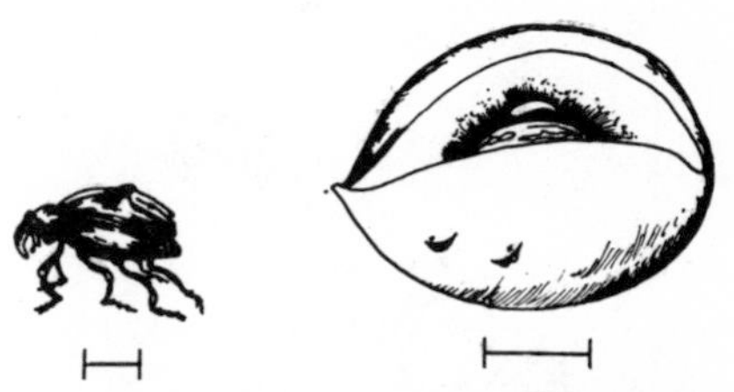

Peach, Cherry and Apple trees, as well as Plum trees are a prey to this tiny determined insect whose adult feeds on tender foliage, buds and blossoms of fruit trees just after they have finished blooming. They cut small circular holes through the skin of newly set fruit. The females cut crescent-shaped punctures as they lay their eggs, and the fruit is thus disfigured. When the egg hatches, the grub is inside the fruit and "eats greedily until it becomes the fat, dirty white worm so well known to fruit growers. When the larva obtains full growth, after some twelve to eighteen days, it bores its way out of the fruit and enters the soil, where it forms an earthen cell in which to pupate."

There are two ways to outwit this insect without resorting to poison. One is to keep hens on the ground directly below the fruit tree. As the larva descends to the soil to continue its life process, the hens see it with their sharp eyes and swallow it forthwith, thus ending the life cycle in a flash. Some people plant plum trees in their henyard to prevent the start of curculio in the first place.

Another method of control is dependent on the curious habit of the curculio to curl up and "play dead" if it is disturbed. It is thus an easy matter to spread sheets on the ground under a tree, then to shake the tree vigorously causing all the adult curculio beetles to fall on the sheet where they may be gathered and burned. This works especially well in the early morning when the beetles are still cold and torpid. The larval stage may be controlled with Dipel*.

cutworm

It is a distressing experience, after one has transplanted a few dozen tomato or cabbage seedlings, to come out the next morning to find many of them cut off at the ground level by the Cutworm. A control suggested in many books is a collar of paper or even a tin can with top and bottom cut out, and the seedling planted in the center so the cutworm cannot reach the stem.

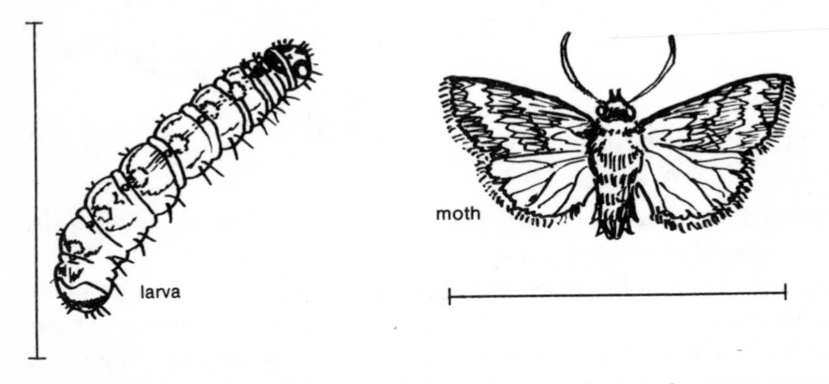

Many years ago someone told us of a much simpler way to outwit the cutworm. When planting the transplant, we stick a toothpick or a matchstick, or even a tough twig or straw directly down the side of the plant stem, touching the stem. When the cutworm attacks the stem to cut it off, he must encircle it, and the tough wood prevents this. This simple method has protected hundreds of young transplants, specifically from this variety of cutworm.

But there is another kind of cutworm which climbs up into the plant to chew the leaves. One year when there was a bad infestation, we used to go out with a flashlight in the evening to hand pick dozens of these monsters. In the home garden it is easy to locate them because they appear first on the calendula, and their eating habits are regular—large irregular areas chewed out, starting from the edge of a leaf. They are not hard to locate in a June evening, because they are large and clumsy and show up well in the rays of the flashlight. The worst part of the job is the touch of the cold clammy cutworm on one's hand. Even this is better than poison bran mash, recommended by some gardeners, which presents a serious hazard to birds and small animals which would also eat it.

A mulch of oak leaves* will also repel cutworms.

If the worst happens and you find favorite seedlings nipped off at ground level, a bit of scratching under the soil surface near the plant will invariably turn up the culprit curled in a ring, sleeping off his feast of the night before.

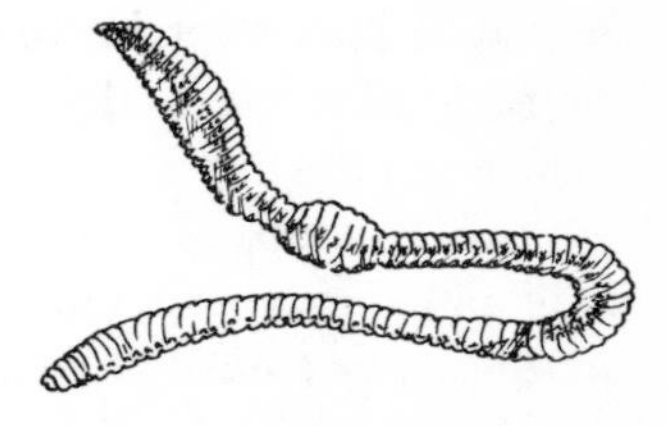

earthworms

In these enlightened days of easily available factual knowledge in great abundance, everyone knows that the Earthworm is not technically an insect, that it belongs to a species all its own, and that it is a particularly valuable living creature, of untold help to gardeners and farmers and conservationists all over the world. We only mention this in passing, with an observation that many books of the Nineteenth Century recommended destroying Earthworms because they were believed to be injurious. Until Darwin's research, people did not realize how valuable the humble earthworm is, nor how unique its contribution to agriculture as it cultivates and adds nutritive elements through the action of its calcareous glands, which are peculiar to the earthworm alone. Earthworm castings make a specially well-balanced food for choice plants and for house plants. In the home garden and on the largest farms if conditions are right, and also in the topsoil of our forests, earthworms are constantly at work, turning and re-turning the soil and keeping it in good condition. Because their process of burrowing involves eating and digesting the soil, and adding their digestive substances before it is expelled, they are a valuable factor in fertilizing and improving soil texture. But *still* there are some magazines which advertise chemicals to kill earthworms in the lawn!

It seems like a waste of money to buy earthworms to import them into a soil that is not built up to receive them, because they may not survive. On the other hand, with good compost, especially using Valerian spray which attracts earthworms and makes their surroundings congenial, there will be plenty of them. Properly made compost will teem with them at a certain stage of its ripening, without a cent being paid out to buy them. No one knows where they come from, but that they do come we have ample evidence in all our compost heaps.

Earthworms can also be raised in shallow boxes filled with soil and with feeding materials. One should make a careful study of all that is known about them, however, because if they are not satisfied, they will leave for better surroundings. On the other hand, if they are moved, even a slight distance, they may return "home" again, since even the earthworm has a strong homing instinct.

Bio-Dynamic Preparation 507 which is made of the blossoms of Valeriana officinalis may be used to keep earthworms where they are wanted. They may be attracted to the compost heap by the addition of onion tops and chicory roots which they relish. The presence of dandelion in any area attracts them and provides tunnels along their deep roots where the earthworms may descend for hibernation.

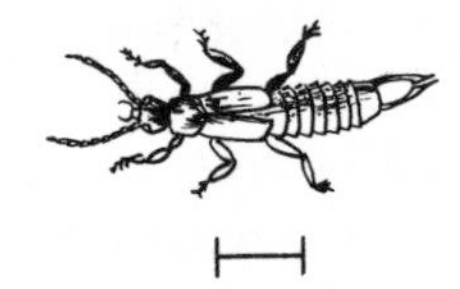

earwig

The Earwig is easy to recognize by the pair of pincers he carries on behind. Sometimes there are large infestations of earwigs that fly at night and sometimes devour many flowers and plants in a short time. They hide during the daytime under rubbish, and there is some evidence that they prey upon other small insects. They especially relish both dahlias and hollyhocks.

They are very easy to trap, and the old horticultural books state that it is possible to rid the garden of every single earwig if one is vigilant in emptying the traps. They like to climb upwards to a place that is high and dry and dark. Hollow stems hung about the garden will be sure to lure them. Another trap is made of a bit of dry moss in the top of a flower pot, inverted and stuck on a stake in the garden.

We know someone who hangs matchboxes from stakes and fences around the flowerbeds. The matchboxes are open just a slit which makes an attractive entrance for the earwig as he goes to rest after his night's feasting. It is easy to empty the earwigs from the matchboxes into a jar of kerosene.

It is most important to empty the traps every morning in order to control the infestation. Fortunately these insects are not regular problems in any one garden. They seem to appear sporadically in subsequent years.

One efficient control is a flock of bantam hens, which add great charm and interest in their happy excursions over the lawn and through the garden. They require a minimum of attention, and even though they peck at a few tomatoes, they do great service in the quantities of insects they hunt and swallow every day.

eelworm or **nematode**

There are thousands of different kinds of Nematodes or Eelworms known to the scientists who have been studying this infinitesimal insect for about sixty years. Of the various species known, some are extremely harmful to certain crops while others may in turn be harmful to other injurious insects. Before one tries to eradicate nematodes from the home garden, one should have a very careful study made to determine if nematodes really are present, and if so, which variety.

Work done in England by the Henry Doubleday Research Association indicates that a material found in some varieties of Marigolds *(Tagetes)* under some conditions in some countries seems to have controlled the potato nematode or eelworm. This research is only in its beginning, and it is to be hoped that in future it will ascertain much more about the eelworm control. It should be mentioned here that a result gained from a certain plant in one country will not necessarily work in another country with a different climate and different kinds of plant life. There is evidence that changing conditions may necessitate discoveries of new substances and processes.

We regret there is no portrait of the eelworm. But the creature is less than one twenty-fifth of an inch in size, and one would not be likely to recognize it from a picture anyway!

flea

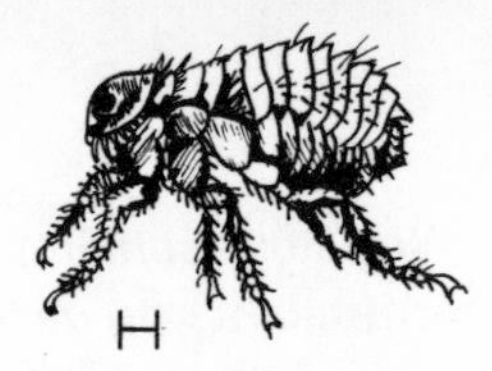

The dog flea needs no introduction, nor does it warrant very much defense. This enlarged illustration explains how the flea can jump so far—as much as thirteen inches! The cat flea differs somewhat from the dog flea. Since it is now generally accepted that DDT and other chlorinated hydrocarbon derivatives are too dangerous to be used on dogs and cats, it is necessary to find less harmful alternatives. Pyrethrum* will stun the fleas and cause them to drop off the animal, but caution should also be used because some people are acutely allergic to pyrethrum.

Bathing or sponging a dog with denatured alcohol or with equal parts denatured alcohol and vinegar or with a strong brew of Wormwood* will be distasteful to fleas. The very common and very aromatic plant Tansy* is also effective in repelling both dog and cat fleas.

For fleas infesting a house, sometimes called sand fleas, try turning on the heat even in midsummer. The dry heat often will expel the fleas without further ado. Another harmless but very effective way is to sprinkle Oil of Lavendar* on the floors and rugs. The sand fleas will disappear *immediately*.

flea beetles

As soon as radish or turnip seed leaves appear, the Flea Beetle hops in with all his relatives and begins to make tiny holes in the leaves. These insects are so tiny that they are practically invisible, partly because they jump away so quickly that when one stoops to examine the plants, they have vanished.

Flying insects are known to dislike moisture, and thus it is possible to control these flying beetles by watering the garden in full sunlight. We have tried both tomato leaves and wormwood branches to repel them and this has worked well in some years. Sometimes the flea beetles have eaten tiny holes in the tomato leaves as well! Soot or lime, sprinkled on the leaves early in the morning when the plants are wet with dew will help repel them. Soot* and ashes mixed in a container also repels flea beetles. Or try laying bruised elderberry leaves over the rows of plants.

Actually the truth is that they do very little damage anyway except to the appearance of the plants. If the plants can be stimulated by Bio-Dynamic Preparation 501 to grow rapidly through the early leaf stage, the beetle damage will be negligible.

Flea beetles may attack tomato leaves at an early stage, but long before the fruit develops the beetles have skipped away and no real harm has been done. This is one insect you can safely identify and then stop worrying about the damage it can do.

fruit flies

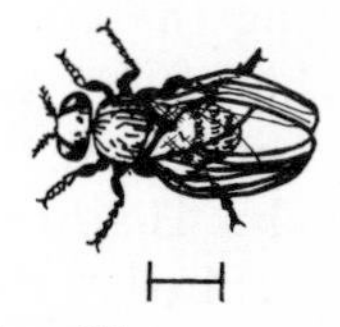

There certainly is little need to illustrate this very familiar summer visitor. Late summer and early fall finds the Fruit Fly in droves around every basket of fruit. The tiny fly does not grow up to be a house fly! It is the adult of its own species and it is developed out of the tiny maggots one finds in decaying fruit in late summer.

Many insect books rather pompously declare that this tiny fly must be controlled with fly sprays, but we have experimented and found a safer method for home use, a method that is much cheaper and easier. All flies are attracted to light. Pull the shades over all but one window. Pull even that shade down until there is about a two inch crack to admit the light. The fruit flies will fly every time to that one light crack. To reinforce the effect of the light, leave a dish of overripe fruit near the window and the fruit flies will gather there by the dozens. Next, pick up the dish of fruit with its covering of flies and rush it out the door while the flies are busy eating the fruit. Once outdoors, the flies will disappear in search of more decaying fruit on vines and trees. They will not return to the house—but there may be more to replace them!

gooseberry caterpillar

Many Americans can no longer boast the luxury of gooseberry and currant bushes in the home garden because of the white pine blister rust which uses these bushes as a host during part of the season. However there is a story which is worth retelling, not as a direct recipe but to demonstrate the principle that merely using a given substance does not always work with the same results. The dynamic interchange of forces also affects the utilization of the substance.

The Head Gardener on an estate in Great Britain about the year 1800 noticed hordes of gooseberry caterpillars devouring the foliage. He dispatched a group of women and boys to pick them off. But the caterpillars were getting ahead of the pickers and were even attacking the currant bushes. As the Head Gardener passed the main house, he noticed a heap of soot, freshly cleaned out of the house chimneys. He also noticed a pile of hardwood ashes from oak and elm. He directed the wood ashes and soot to be mixed together and applied to the gooseberry bushes while the dew was still on them in the morning. This was done and the caterpillars were promptly overcome.

Pleased with his success, he made sure that soot and ashes were stored in piles on the ground in a shed, ready for future use. Year after year he tried the combination to control the caterpillars, but he never was successful in immediately quelling a large infestation until—and when he began to think of Process and Forces instead of simple Substances—twenty-five years later. While trying to puzzle out why it was no longer effective, he recalled that the original soot and the original ashes had been freshly removed from the chimneys and fireplaces—the day before they were used. In later years, both soot and ashes had been stored on the dirt floor of the shed. The exchange between soot and soil on that dirt floor had leached out something.

He sent for fresh soot and fresh ashes and lo, the effect was immediate: the gooseberry caterpillars gave up under the powdering of soot and again were vanquished, thus proving the Head Gardener's point that both materials to be efficacious must be used fresh.

He adds that soot alone will destroy the caterpillars, but the addition of wood ashes makes it adhere better to the leaves, and prevents the mixture from flying about. He also adds that wood ashes used in damp weather in full strength will kill and clean any kind of moss or lichen off trees. This is the only substance he ever used for that purpose.

I quote this story, not necessarily to be followed exactly but to direct attention to the interaction between fresh soot* and fresh ashes and early morning dew which all worked together with the

English climate in 1814 to kill the caterpillars. It might work in America this year or next, but even if the same *substances* do not work here, the *principles* should still work, and it is up to us to use imagination and ingenuity and observation to find our own remedies to replace the much more dangerous agricultural poisons which we are trying to avoid.

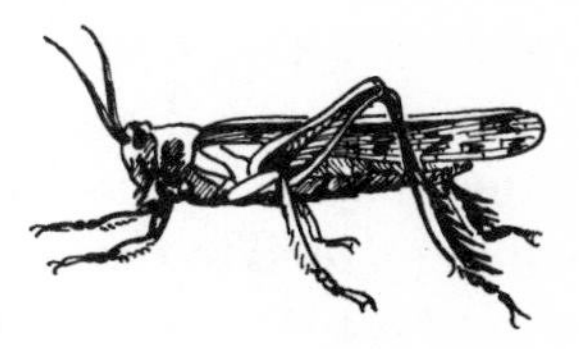

grasshopper

No one who has ever lived through a scourge of grasshoppers can ever forget the long, hot summer days when every vestige of green gradually disappeared first from the herb garden, then the vegetable garden and orchard and finally from the corn fields, too. We never would have believed it possible actually to see the green of the corn leaves diminishing before our very eyes. It finally was decided to make the corn into silage immediately (even though it was immature), before the grasshoppers ate every leaf, tassel and stalk. (Grasshoppers that got into houses even ate the draperies at the windows.)

Three years in succession this happened, and we came to realize how devastating a single species of insect may be to an economic situation. The one redeeming feature was that we had plenty of grasshoppers to study and to use for experiments.

The first year we made brews of everything we could imagine that might repel the hoppers! Soap suds, rue extract, wormwood (they had already consumed the wormwood plants), castor oil, urine, lime, wood ashes and tar. Nothing stopped them. We studied the local plants they left untouched—there were only two or three species like hoarhound and castor bean. We made brews and sprays of them, hoping to find a natural repellent, but to no avail.

The second year we cultivated a large flock of hens and started early in the spring when the hoppers were small. The hens enjoyed the feast and followed wherever we went, snatching up the young grasshoppers as we knocked them down with a long bamboo pole. This made a noticeable difference to the grasshopper damage to the gardens. In orchards there was no such control and the hoppers ate every leaf and every peach, and they hung on the bare peach stones which hung from the bare branches.

The third year we hoped their cycle would be finished, but they arrived in the early spring, tiny facsimiles of last year's horrors. The hot, dry weather was far from discouraging to them, and again we had to use flocks of hens and bantams to keep them controlled. As the season wore on and the grasshoppers grew larger, the hens could eat only two or three at a time. Finally they turned against them altogether, and we turned to government bulletins which told how the peoples of the Orient have used several species of insects for human food in times of need.

One day while examining a grasshopper through a magnifying glass, we noticed a small red speck under its wing. Another hopper revealed several such specks. A government bulletin printed in South Dakota described this as the Grasshopper Mite or Grasshopper Tick as it is also called. This mite lives through the winter in the soil. When the weather warms up in the spring, the mites search out masses of grasshopper eggs and gorge themselves on the contents. The female mites lay their eggs in the soil where the larvae hatch out as tiny six legged mites. Eventually they attach themselves to the body of a grasshopper where they suck its fluid. There may be as many as 35 to 175 larval mites on one grasshopper. They attach themselves to the thinner, softer parts of the grasshopper and weaken the adult. Perhaps in time the Grasshopper Mite may be developed as biologic control for this very serious insect pest.

The USDA Book of Agriculture for 1952, INSECTS, tells about an old-time method of collecting the grasshoppers to remove them from the fields. It is called a hopperdozer, a long narrow trough of boards so constructed that it will run broadside over the fields causing the hoppers to jump up against a board, which knocks them back into a trough filled with water with a small amount of kerosene floating on the top. The government entomologist reports that up to 8 bushels of grasshoppers to the acre have been collected in this way. This same procedure might be modified to catch other insects invading in large numbers.

It should be noted that at any time during our three summers it would have been possible to use strong poisons to rid ourselves entirely of grasshoppers, but we considered this a singularly good opportunity to experiment to find safer methods of control. It also was a challenge to see how much we could salvage from garden and orchard before and after the grasshopper scourge. That was the year we had to make horehound candy for Christmas gifts because every other plant in the entire homestead had been stripped.

From our viewpoint as human beings, in spite of the grasshoppers' harm, it was impossible to hate or be angry with them. Who could feel hostility toward such a ludicrous figure with its gloomy eye and its horse-like face, propelled by gangling rose-colored hind legs?

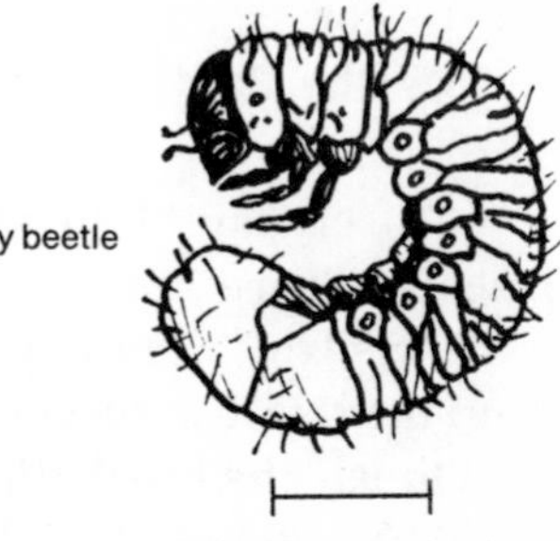

GRUB, may beetle

grubs, in general

The home gardener meets up with various kinds of Grubs or larvae as they are technically called. There is the fat white grub with a tan spot on its tail, which will mature into the June Beetle. There is the Japanese Beetle Grub, cosily developing in his underground cell in the earth. The grub of the honey bee develops in the wax cell in the honeycomb where the queen bee laid her eggs. Worms and caterpillars are also larvae of a similar type that have made adjustments to live out in the open above ground. Domestic poultry allowed to run and scratch in the fall-plowed garden will help keep these grubs to a minimum.

Not all grubs are injurious, although we hear the most about the ones that do damage in the garden. The worm and the caterpillar and the maggot are also larval stages of various insects in the first step after the egg hatches. This is the tender, juicy defenceless stage which the birds can catch and eat with great relish, thus assisting in insect control. Even though some birds may not eat a full-grown beetle because it is too crusty or gives off a protective smell, they may catch myriads of grubs of the same kind of insect to feed their young. Naturalists who have studied and recorded the habits of birds state that growing birds require almost their own weight in insects every day. The parent birds of many species, especially the starling, cover all possible branches and twigs from the earth up to the treetops, in constant motion and search to find enough food for their hungry young. It is worth noting over and over again that in the deeply-planned economy of nature, the right things happen at the right time:

Baby birds hatch out when the weather is growing warmer, which is also the exact time that the insect worms and grubs are also hatching out in the warmth. The insect grubs are there in time to feed the baby birds. The bird world and the insect world can be seen at work in close cooperation, if man has patience and insight to learn.

gypsy moth

The Gypsy Moth has caused defoliation of widespread areas of forests. There are various methods which are effective for small scale control. Scrape off the large, easily visible egg clusters and drop them into kerosene or burn them. Egg clusters can be destroyed by a touch of a brush dipped in creosote.

The wingless female Gypsy Moth cannot fly. Therefore, a band of tanglefoot* or of grease* will prevent her from climbing the tree trunks. Hay bands* wrapped around the trunks in May will provide an attractive dwelling place where many caterpillars will congregate all summer to live. They may be collected and destroyed at the end of the caterpillar season, which will vary in different regions.

Bacillus Thuringiensis* (BT) is a bacterial insecticide applied by spray which is aimed at caterpillars of moths and butterflies. It is, however, not as effective on Gypsy Moths as on a number of other pests.

Another biological control is the Japanese insect, Oercyrtus Kuvamai, now produced by USDA laboratories and released in New York, New Jersey, and Pennsylvania. The female of the parasitic fly lays her eggs in the eggs of the Gypsy Moth destroying 30-40 percent of them.

Dispalure or Gyptol is the name of a synthesized sex attractant used to lure male Gypsy Moths, which can then be destroyed.

harlequin bug

This is an unusually handsome insect of the true bug family. It is related to the squash bug and all the other so-called shield bugs, which are so called because of their shield-shape. It is easy to recognize because of its attractive color scheme of orange and yellow and black and white. The eggs are laid in plain sight and look like rows of tiny kegs placed in neat rows on the plant leaves.

On our Missouri homestead we used to have considerable trouble raising fall turnip crops, partly because the summer weather was so hot that seeds failed to germinate and partly because Harlequin Bugs moved in and ate every turnip seedling down to the ground, precisely. Later we learned that one method of control for harlequin bugs is to plant a few rows of turnips or mustard greens or cabbage for what is called "trap crops."* The insects like these plants so much better than any other plants, that they can be trapped there and brought under control to save other plants in the garden. In other words, we had planted a trap crop accidentally—and we certainly trapped a fine crop of harlequin bugs.

honey bees and **wasps**

Literally thousands of books have been written about the honey bee. There are, however, still more thousands of bee keepers who have never had to read books about bees. We only mention the honey bee here in the hope that we can differentiate between it and wasps, for many times honey bees are blamed for mischief done by wasps.

In general the honey bee is smaller than a wasp, she is golden in color with a somewhat fuzzy look while a wasp is yellow or black and much harder looking. The honey bees tend to be gentler in disposition and usually will not sting unless they are caught in one's hair or clothing. The honey bee can sting only once and her life is

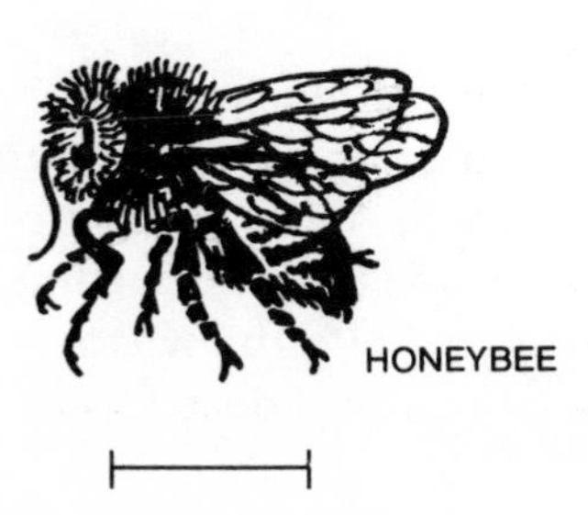

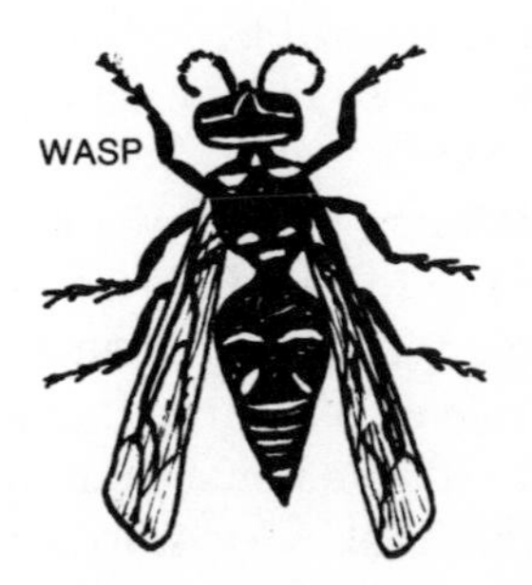

ended. They work hard all summer, diligently making honey and rearing their young brood.

The wasps on the other hand have a waspish disposition and will pursue and sting many times over if disturbed. Wasps, or hornets as they are also called, build nests of mud or of chewed up wood which they manufacture into paper.

We once had a wasps' nest outside a window of the sunporch. It was perfectly safe to stand inside, close to the glass, with two magnifying glasses held in focus so we could watch each wasp worker as she flew in with a load of cellulose "stuff" to build the nest. The encyclopedia told us that the cellulose comes from dead trees or fence posts or unpainted lumber which the wasp chews. She has some way, not yet understood, of collecting, wetting and transporting the building material. In minutest detail we watched each wasp add her new load of wet paper-stuff to the edge of the cell already constructed. The new layer was less than one thirty second of an inch wide, but it was plastered in a regular strip along the edge of the last layer, while the wasp smoothed it by drawing her jaws along, one on either side. One load was enough to plaster a strip about three quarters of an inch long, and about one sixty-fourth of an inch thick.

When the sun shone we could see right through the papery walls of the wasps' nest. If the wind blew we could see the paper tremble, and yet it was strong enough to shelter a large colony, of perhaps ten thousand hornets.

Naturally we were careful not to upset the wasps by slamming the door, nor to stand in front of the nest when the weather was damp and threatening or cold and blustery. Weather conditions have a strong effect on insects, and a coming storm will make even the gentle honey bee tense and angry. That season we agreed not to use the sunporch door, preferring to let the wasps reign supreme in the neighborhood of their paper citadel. On the one occasion when they did sting, we rubbed the places with cut onion and the pain disappeared. We kept onions handy the rest of the season in case of sudden need.

Evidently others were also watching the wasps' nest. Before frost in the fall had slowed down insect life for good, we noticed a large hole torn near the entrance to the wasps' nest. As winter wore on the hole was enlarged. One day in the dead of winter we found out what did it. The house sparrows were digging into the wasp nest and eating the fresh wasps preserved there in a natural deep freeze.

Female wasps make their appearance in April and May. Each one is the origin and mother of a wasp colony. If the first ones are dispatched, the colonies never get started. A fly swatter is a good, sure, final method of dispatch! It is not very safe to burn out a wasps' nest. Many houses have been set on fire thus. Why take a chance when winter will finish them off?

The whole family of wasps is carnivorous, and eats other insects, especially caterpillar (including cabbage worms), and beetle larvae. The female wasp packs the cells of her nest with paralyzed spiders or insects. Then she lays an egg in each cell. As the wasp larva emerges, the paralyzed insect becomes its food supply.

A nest of yellow jackets is something to treat with great respect because they will sting at the slightest provocation and will continue to sting over and over again. Since they sometimes build nests in the ground, they may be stirred up by a plow or a lawnmower. Mud is often the quickest first aid for hornet stings, but sometimes they are serious enough to necessitate help from a physician. One may be stung also by a bumblebee, whose home in the ground has been disturbed.

If you want to witness a battle between two insect species, keep watch for the time when a wasp or a hornet ventures too near a spider's web. The two are about equally matched in prowess: the wasp is more aggressive but the spider is subtler. The spider will wait until the wasp is well entangled in the cobweb, until he cannot fly away, and then the spider will try to bite and suck the wasp's vital fluids. Since both creatures are both our friends and our foes, sometimes one, sometimes the other, it is a puzzle to know which to applaud. Sometimes the fight will go on for an hour or more. If you have time to watch to the victorious finale, you will come away impressed with the beauty and skills that have been created in the insect kingdom. Technically speaking, of course, the spider does not quite belong in the insect kingdom.

housefly

The common Housefly is too well known and too unpopular to be pictured here! Everyone has a pet remedy for houseflies all the way from the twenty-cent fly swatter (which still stands supreme) to an expensive trap which electrocutes any fly that touches it. See

Electronic Insect Trap.* There are various sprays said to be harmless to human beings and to animal pets, but even these should be used with the utmost care.

Some people believe that a large wad of absorbent cotton attached to a screen door, usually tied on in the middle of the screen, will frighten the flies away because it resembles a huge moth. Other people prefer mechanical methods of trapping houseflies either with an old fashioned wire screen trap with bait inside, or in a "Big Stinky fly trap," available from Mother's General Store Catalogue, Box 506, Flat Rock, N.C. 28731.

A stream of cold air jetting across the doorway will keep flies from entering a barn door. Flies may be attracted to a small slit of light at a curtained window where it is easy to kill them with a fly swatter in the early morning when they are still slow and torpid from the night's cold.

One fact has been proved: poison sprays like DDT only succeeded in creating an immune strain of flies. Some of the hardier individuals escaped the poison and went on breeding DDT-resistant strains of flies. Their progeny continued to be immune to the old poison which now had to be replaced by a stronger poison. In the years that DDT has been used, we have all had a chance to observe this, even though they seemed to abate for a few years. It is to be hoped that scientists will not go on inventing stronger poisons, but that they may again try other less-toxic materials to keep the housefly under control. DDT is now theoretically outlawed but has been replaced by other chemicals.

hover fly, syrphid fly, tachinid fly, sweat bee

The Syrphid Fly belongs to a large family which in some places is called a Hover Fly because of its habit of hovering over the flowers. These Syrphid Flies are helpful predators and their larvae feed on aphis and scale insects to such an extent that they are of considerable economic importance to fruit growers. The Syrphid Fly family has only two wings, and no equipment for stinging. It is harmless to human beings and altogether a most helpful, prolific and desirable insect to cultivate.

Closely related to the Syrphids are the Tachinid Flies which also have only two wings and which are most numerous and active. The

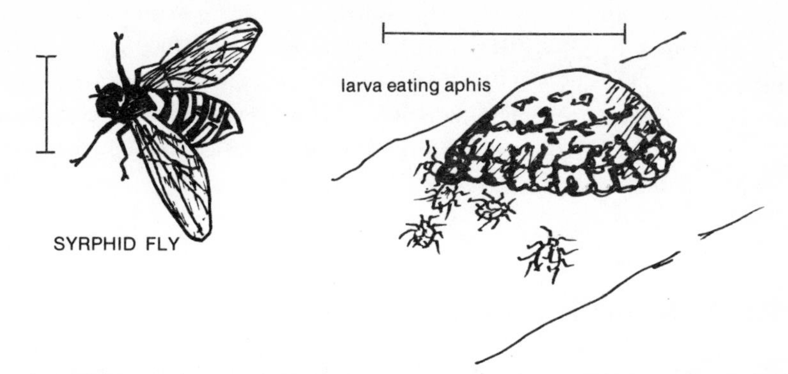

distinguishing name, Tachinid, refers to the swiftness with which they can fly. They also are of economic importance because they are parasitic upon many different species of caterpillars. One species helps control Gypsy Moths by laying eggs so that the larvae can devour the bodies of the Gypsy Caterpillar. Each tachinid fly has a different way of locating its grubs in the host insect. One fly deposits living larvae in tiny cup-shaped receptacles where the larvae can intercept the silken threads upon which Brown-tail caterpillars travel to and from their nest. As the Brown-tail returns to its nest, the larva attaches itself to it, to the ultimate destruction of the caterpillar.

The tachinid flies look to the amateur like ordinary house flies, except that they do not come into the house but work diligently outdoors.

Another insect which hovers around human beings is the Sweat Bee. It sometimes is confused with the hover fly, but has four wings and can inflict a slight sting. He is otherwise relatively small and insignificant compared to the great helpfulness of the syrphid flies and the tachinid flies, which both deserve all the encouragement we can give them—and their tens of hundreds of relatives.

ichneumon fly

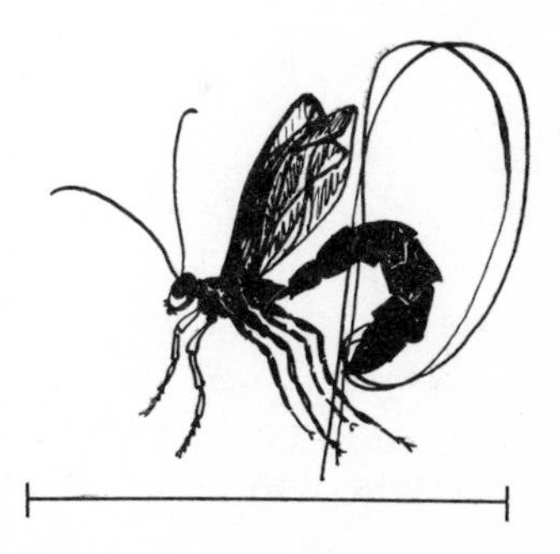

Loudon wrote in 1824: "Nature has furnished a remarkable insect which assists man in the destruction of the caterpillar, the Ichneumon Fly. The insects of this genus lay their eggs in the bodies of caterpillars or pupae, which are there hatched; the larvae have no feet; they are soft and cylindrical, and feed on the substance of the caterpillar, which never turns into a perfect insect, while the larvae of the Ichneumon spin themselves a silky web and change into a *pupa incompleta*, and in a few days the fly appears."

japanese beetle

The Yearbook of the U.S. Department of Agriculture, 1952, gives excellent information in color about the life cycle of the Japanese Beetle. The adult female lays eggs in the soil. The eggs hatch into a grub which winters in the soil. In spring the full grown grub eats grass roots until it is ready to pupate in the soil. The pupa develops underground until the adult beetle emerges. The underground life of the Japanese Beetle takes about ten months of the year. In May or June the adult emerges from the earth and goes straight to its favorite plant: smart weed or grape vine or rosebush or apple tree.

Another favorite plant is white geranium where the insect can be found and destroyed in a small can of kerosene. Experiences differ as to the insect's behaviour. Some say it will fall to the ground unconscious. Others report that the beetle only comes to eat or to rest on the white geranium which acts as a bait plant. Bright yellow beetle traps baited with geranol and ergenol will also lure them.

The Yearbook carries farther information that has been tried and is recommended by the Department of Agriculture. See Doom* or Japanese Beetle Milky Spore Disease* in the back of this book. People who have used Doom report it to be effective to a high degree. Some communities have made it available to their residents with good effect in controlling the beetle on a large scale. This is a good example of a biological control which deserves our enthusiastic support in all its phases.

Many people who have not used Milky Spore Disease, but who have had some experience with Japanese Beetles report that during the ten months of the year when the beetles are in the grub stage underground, the birds have worked hard over their land and have controlled the beetle. There is evidence that some birds eat the beetle in its adult stage, and it is easy to see how the fat white grub of the beetle would be a juicy morsel for the bird equipped with long beak to burrow into the surface of the soil. Flickers are most active burrowers, but the starlings arrive in great numbers and carefully explore every inch of soil for the grubs. Nor is it true that birds would rather eat grain and fruit than these grubs. First-hand reports tell us that when the insects are there, the birds abandon well-stocked feeders to go and gather grubs to feed their young. Birds do not carry grain or bread crumbs to their offspring, but each spring one sees many birds carrying worms and grubs and beetle larvae to their hungry broods.

Last summer we had a small field of wild primrose which we left for a trap crop for Japanese beetles. The yellow blossoms were systematically eaten. Late in August we noticed yellow blooms returning, and upon investigation we found large black and yellow garden spiders with webs spread about 18 inches apart all over the field. The spiders were voraciously devouring every Japanese beetle caught in their webs. In a few seconds the spider would wrap and bind up the beetle in web and then eat it. We don't know how many Japanese beetles were destroyed because all the debris fell to the ground under the spiders' webs. This year we again will encourage the wild primroses and welcome the big black and yellow spiders.

These is some evidence to indicate that planting Angel's Trumpet in a garden keeps away the Japanese Beetle. This has been reported by several eye-witnesses and would be a suitable subject for

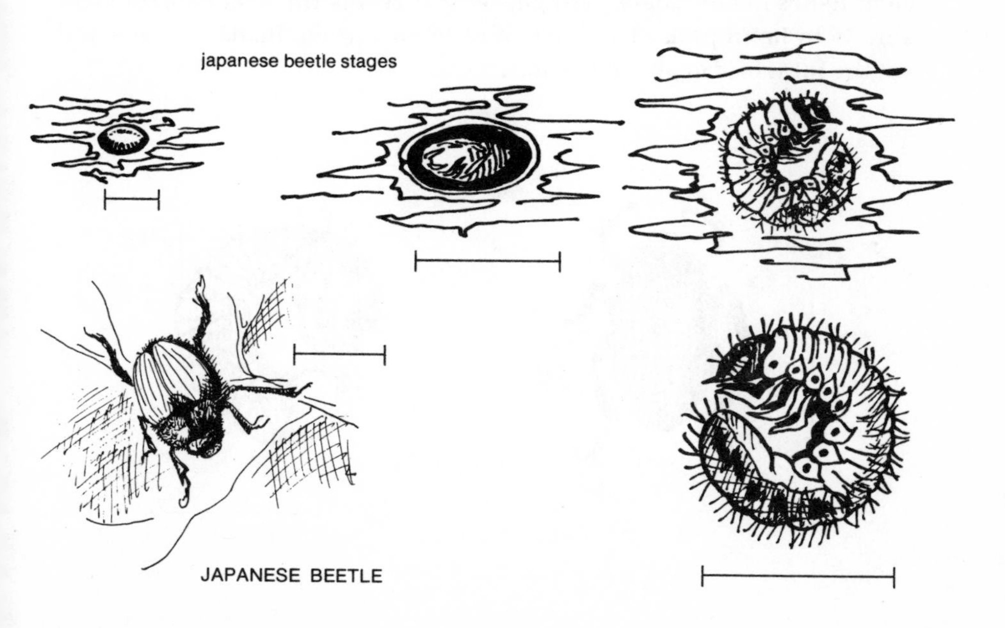

experiment and research. One should be careful with Angel's Trumpet, however, because it is really Jimson Weed (Stramonium) and is a strong poison plant. The seeds are especially tempting for children to put in their mouths because of their attractive shape and size.

Japanese beetle traps, colored bright yellow and used with bait, may be purchased from seed houses or hardware stores.

june bug

The June Bug, which more properly should be called the May Beetle is the very common insect that bumps into automobile headlights and flies round and round outdoor lights in May and June. In some parts of the country it seems to do more damage than in others. Its larval stage is spent as a fat white grub in the earth, and it is relished by the birds. The adult beetle often can be found in the garden after dark, eating its favorite plants. We have found them feasting on the petals of rosebuds. It is such a large insect that it can eat half a rosebud at one sitting. Handpicking the two or three that arrived in early evening seemed control enough, because none flew in after that to continue the night destruction.

Considering the small amount of actual damage the june bug accomplishes in most backyard gardens, it seems the best control measure is to hand pick after dark, and to encourage birds to come and eat the fat white grubs in the soil.

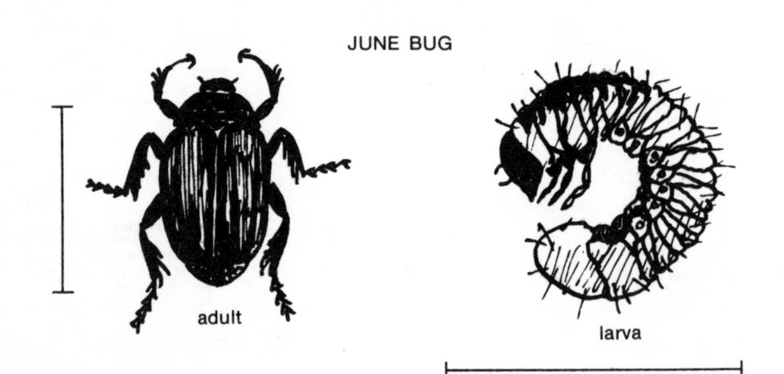

lady bug

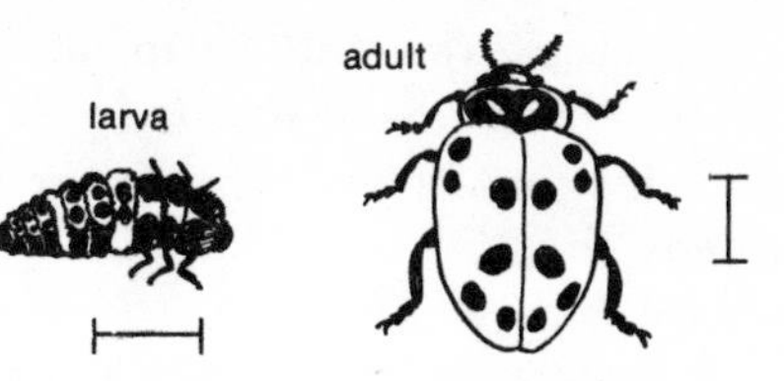

The jewel-like scarlet beetle which shows up in bright contrast to the green leaves in the garden patch really should be called Lady Beetle because it belongs to the beetle family rather than to the bug family. Sometimes it is called the Lady Bird Beetle. The name was originally Our Lady's Beetle because hundreds of years ago people recognized this insect to be helpful in eating the less desirable insects in the garden.

The lady bug deserves protection because she is such a help in controlling aphis. Whatever insecticides are used to control other insects, even the bad bugs, will kill the lady bugs too. For this reason, many home gardeners prefer to hand pick the bad insects or to check them by other means: baits, traps or repellents, rather than to take any chance of losing the lady bugs.

Because the lady bug larva looks so different from the adult, we want to call special attention to its uncomely appearance. In spite of its lack of beauty, it is a true friend to the gardener because it uses aphis for food. In this larval form the lady bug consumes many aphis every day. See more about the lady bug under Aphis.

leaf miners

Many different plants are attacked by Leaf Miners whose work begins to show on spinach, Swiss chard, beets, columbine, plantain, birch and many other plants around the second or third week in June. It is often quite possible to see the small grubs between upper and lower skins of the leaf. The leaf miner is inside the leaf, beyond the reach of anything but the strongest poisons. For the small home garden it is quite practical to gather all infested leaves by hand and burn them. For a small birch tree, for instance, it is not difficult to pinch each infested leaf one notices, to crush the grub inside. This kind of insect, after tunnelling through the leaves of its choice, will drop into the ground and form a tough pupal case where it will spend the next stage of its development. Wild birds help control it in the ground if the soil is well cultivated. Strong smelling herbs or tar might repel the adult fly in early spring so she would not lay her eggs in the birch tree or on the columbine plant. Leaf miner damage is not serious and is one which merits one of these minor kinds of treatment.

leaf rollers and **leaf tiers**

Sometimes one finds abnormal conditions in the home garden such as a small twist of leaves with an insect, usually some kind of worm or caterpillar, inside. This is extremely simple to control at this stage. One has only to pick the distorted leaves and burn them. Do not throw them into the compost heap because the insect will escape and start all over again. Hand-picking is a cheap and efficient method of control for many kinds of insects in the small backyard garden.

mealybug

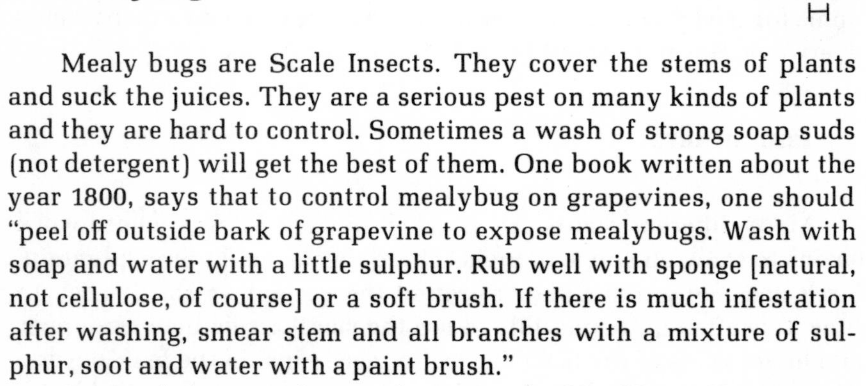

Mealy bugs are Scale Insects. They cover the stems of plants and suck the juices. They are a serious pest on many kinds of plants and they are hard to control. Sometimes a wash of strong soap suds (not detergent) will get the best of them. One book written about the year 1800, says that to control mealybug on grapevines, one should "peel off outside bark of grapevine to expose mealybugs. Wash with soap and water with a little sulphur. Rub well with sponge [natural, not cellulose, of course] or a soft brush. If there is much infestation after washing, smear stem and all branches with a mixture of sulphur, soot and water with a paint brush."

For mealybugs on house plants try the Hot Water Treatment*. Another simple home practice is to use a cotton swab dipped in Denatured Alcohol* and touch each mealybug in turn. The alcohol penetrates the waxy protective covering and the mealybug will expire. This is a rather impossible treatment for a large orchard, however!

In fruit trees, if soap suds do not control the mealybugs, and if the situation is desperate, Kerosene Emulsion might be used*. It must be remembered, however, that a strong insecticide like this would also kill lady bugs and other beneficial insects. It was because of the mealybug in California that the Australian lady beetles were originally imported. The USDA Yearbook for 1952, INSECTS, states "The citrus mealybug and the citrophilus mealybug were satisfactorily controlled by this means for many years. State, county and private organizations were engaged in producing the beetle and hundreds of millions were reared and released at a cost of approximately $2.50 a thousand. The need for this program became less acute in the 1930's because of the introduction of highly effective internal parasites in several of the mealybug species."

mexican bean beetle

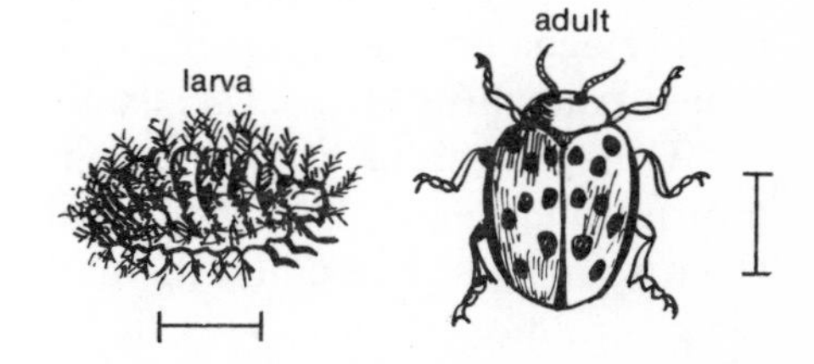

The Mexican Bean Beetle is the only member of the lady bug family that has habits distasteful to gardeners. Even Entymologist Lutz, who usually has good things to say about the worst insect, writes: "It should be disowned by its family."

However to the backyard gardener the Mexican bean beetle is one of the easiest to control. The method is to plant a row of potatoes between every two rows of green beans (not lima beans). There is something about the potato plant that repels the Mexican bean beetle and something about the green bean plant that repels the Colorado potato bug. No one quite understands why this is, but thirty years of practice makes us personal witnesses to the efficacy of the method.

This illustration shows the characteristic spots on the beetle's back which indicate it to be a member of the lady bug family. It is also of some help to observe that the larva also wears the same spots as the adult beetle. Both the beetle and the larva are a rather unappealing shade of grayish yellow color.

mosquito

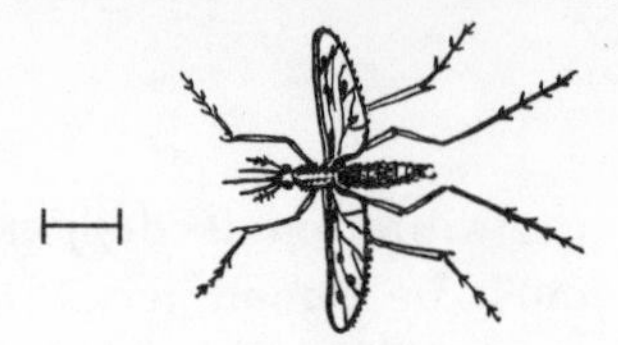

Who needs a picture of the Mosquito? Most of us would prefer not to see even a picture of this real pest. In addition to being most annoying, the mosquito has been responsible for many thousands of cases of malaria and other serious diseases. As one reads through the records of the U.S. Insect Control history, one finds more frequent reference to mosquito control than to any other insect. It is clear to see that mosquito control on a large scale is very important both from a medical and from an economic standpoint.

Probably there are more home repellents for mosquitoes than for any other common insect pests: oil of citronella or oil of pennyroyal rubbed on the skin will keep them from biting. Castor bean plants near a door are said to keep mosquitoes away. Tansy will repel them. Rain barrels and places filled with stagnant water, unless furnished with goldfish, should be covered so the adult mosquito will not lay eggs on the water. Electrocuting night lamps will kill quantities of mosquitoes. See Electronic Insect Trap*.

Others besides human beings are doing their best to keep the mosquito under control. Hawks and swallows and whipoorwills eat them as fast as they catch them. The purple martin is said to eat its weight in mosquitoes every day. One insect is even called the mosquito hawk—the dragonfly—because it eats so many mosquitoes as it darts over lawns and lakes in the evening twilight. Any insecticides aimed at killing the mosquitoes may also kill off the predatory dragonflies unless scientists develop selective insecticides. Still another helper is the bat which eats quantities of mosquitoes and other insects during his evening flight.

If someone should ask "What good are mosquitoes anyway?" the answer might be; "To furnish food to all these predators that consider them such tender morsels."

It might be added here that with the progress of civilization away from the countryside into more urban living, many of our insects seem useless and ill-designed to fit our present patterns of living. This of course is true. Man has changed, but the insects and their relation to nature have not changed in centuries. In 1824 Loudon wrote: "In southern regions there is a larger species (of gnat) which is known by the name of mosquito. Its bite is painful, raising a considerable degree of inflammation, and its continual piping note is exceedingly irksome where it abounds, especially during the night. When it settles to inflict the wound and draw the blood, it raises its hind pair of feet. In Lapland the injuries the inhabitants sustain from it are amply repaid by the vast numbers of water-fowl and wild-fowl which it attracts, as it forms the favorite food of their young."

Here is an answer to the question "What good are mosquitoes anyway?" It is quite possible that the experimenters of the future may discover certain food elements in the body of the mosquito which are essentials for water-fowl and for fish, just as specialists in nutrition have learned about vitamins and minerals which are indispensable for human nutrition. . . . "There's more in heaven and earth than is dreamt of in any philosophy". . . that is, there is more in real life and in the dynamic relationships between living creatures than anyone could find in book learning.

MOTHS IN FRUIT TREES

moths in fruit trees

The general principle that Moths respond to strong odors and are repelled by some, may be applied to the moth stage of certain insects that do damage to fruit trees. Some of the aromatic odors are camphor*, tar* and the Aromatic Herbs*. For instance, for many years it has been common practice to plant tansy near the peach trees to repel fruit moths. Other moth-repellent herbs, which also work just as well in one's winter woolens when they are stored, are pennyroyal, wormwood, old man-old woman, sage, santolina, lavendar and the whole mint family.

See also in *Recipes and Formulas* Baits for Fruit Tree Moths, Cotton Batting Band around tree trunk, Grease Band around tree trunk, Trap of wire netting for Canker Worm, Electronic Insect Traps and Hay Band around tree trunk.

onion fly and **onion maggot**

There is a small Maggot which infests onions, hatching in the onion bulb just below the surface of the soil. Several generations of these maggots may be stored in onions during the winter. They completely destroy the onion inside its papery outer layer. Some gardeners recommend wood ashes scattered on the surface of the soil to repel the onion fly, which lays the eggs in the onions in the first place.

The Onion Fly looks so much like an ordinary house fly that even experts must observe it closely to tell the difference. Any aromatic substance which will keep the onion fly away will help save the onion crop. Some home gardeners watch for the root maggot in the onion bed, and if any plant shows signs of dying they pull out that plant and burn it quickly. It is most important to keep the onions growing rapidly to keep ahead of the root maggots. Therefore a special dressing of compost and frequent watering should be beneficial to the crop.

pillbug or **sow bug** or **wood louse**

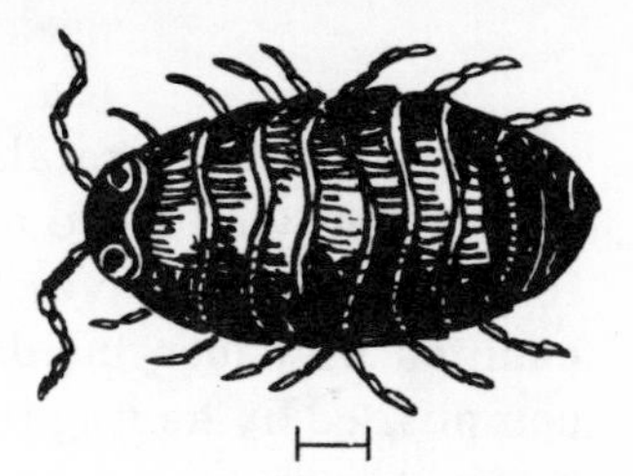

Few of our common garden creatures can boast as many or as varied names as the Pill Bug, which really is not an insect at all. In technical books the pill bug or sow bug is found in the section which includes lobsters and spiders and horseshoe crabs. Adult insects never have more than six legs, the spider has eight legs, but the pill bug has at least fourteen legs! However the behavior of the pill bug resembles that of the insects, and perhaps that is why they are often called Wood Lice.

Sutton wrote many years ago: "Wood Lice are terrible destroyers, but are easily caught and may be completely eradicated by perseverance. When a frame or pit is infested they may be destroyed wholesale by pouring boiling water down next the brickwork or the woodwork in the middle of the day. If this procedure does not make a clearance, recourse must be had to *trapping*. In common with Earwigs, they love dryness, darkness and a snug retreat: But as a mere home suffices for Earwigs, a home with food is demanded by Woodlice. *Take a small pot,* quite dry and clean. In it place a fresh-cut slice of potato or apple or lettuce or spinach, fill up with dry moss, and turn the whole thing over on a bed in a frame [cold frame] or pit [pit greenhouse]. Thus you have devised a Woodlouse Trap, and next morning you may knock the vermin out of it into a vessel full of hot water, or adopt any other mode of killing that may be convenient. Fifty traps may be prepared in a hundred minutes and those who are determined to get rid of Woodlice may soon make an end of them."

When a Sowbug is disturbed, sometimes it will roll itself into a perfect ball, covered on the outside by its jointed shell. This explains its other name, "pillbug".

Unless it is absolutely necessary, perhaps it is just as well not to kill off all the pillbugs, because there must be some reason why they exist among growing plants. In our greenhouse we found that they were always abundant when conditions were getting too dry. They were also abundant in compost which was made mostly of goat manure. In a more moist climate, with compost made with cow manure, we have found pillbugs only in their normal habitat—under a large rock or beneath logs in waste places.

While they were abundant in the greenhouse, we used to find them when we turned on the lights at night, greedily devouring the tiniest seedling leaves. We lost a few seedlings, but even so, we counted that loss incidental to the good that might have been accomplished by having the entire gamut of living creatures present in a greenhouse.

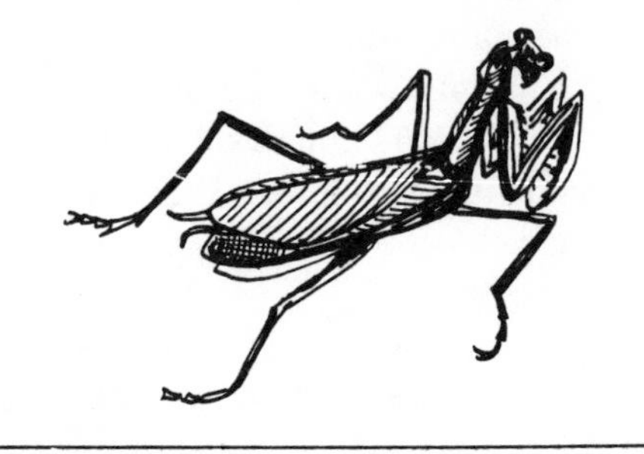

preying mantis

The Preying Mantis is so spectacular that he, or rather she, (because the female devours the male at a certain stage of their life cycle) has always received a good deal of enviable publicity.

One sees them whirling around the corner of the house at dusk to take up their ambush on a particular branch of a certain shrub where the same mantis may be located night after night awaiting her prey. Of all the fascinating figures to be encountered in the insect world, probably the preying mantis is the most startling as well as the most fantastic. With her head turned almost completely around backwards, she seems to stare at you with big green globular eyes, and you feel yourself almost to be her prey.

We used to enjoy watching the preying mantis chew up any number of insects, even our own honey bees, holding them in her front "paws" and gouging out bites as though she were eating an ice cream cone. For every good honey bee she ate, we estimated she ate many more injurious insects.

Naturally, if one is fortunate enough to have a few preying mantises in one's backyard garden, one would not risk their safety by using any kind of poisonous insecticides on any other insects—even on mosquitoes. If the preying mantis is given an opportunity, she will eat many mosquitoes every evening.

Preying mantises can be purchased from Eastern Biological Control Co., Route 5 Box 379 Jackson, N.J. 08527 or from garden supply and seed houses.

raspberry cane borer

A slender beetle with black wing covers and yellow thorax cuts two rings of punctures about an inch apart around a young shoot. Here she lays an egg which hatches into a white grub which begins to eat the pith inside the cane. The top of the cane fades and dies. The control is reasonable and not difficult! Cut the wilted shoots well below the injury and burn them.

red spider mite

Every insect book carefully explains that the Red Spider Mite is not a true insect, and that it belongs more closely to the family of the horseshoe crab than to any of the insect families. Be that as it may, the red spider mite is the plant pest *par excellence.* The worst aspect is that it is so tiny it is practically invisible, and therefore gets a head start before anyone suspects it is at work on any plant. If plants look unthrifty and with tiny tents of fine cobweb on terminal leaves, look with a magnifying glass on the underside of the leaves. With a strong glass and good light you may see tiny specks about the size of fine meal. Magnify these fifteen times and they will look like the drawings—a very pale reddish color, with eight hairy legs.

Red spider likes a dry atmosphere. If the atmosphere can be kept humid the battle is half won. Syringing the plants or washing them with a hose with fine spray under pressure will help dispose of this tiny pest. Be careful not to get soil too wet for the plants. Loudon wrote: "Neither should it be performed [spraying with water] when the sun acts freely upon the plants lest their leaves become disfigured, for the water forms itself into spherules and thereby the leaves are disfigured by being burned in the focus of each spherule."

Forsyth recommended watering infested trees with clear Limewater* over the leaves to control red spider mites. Nettle Brew* plus Equisetum Tea* are effective also. On large-scale fruit plantations red spider is controlled by a thin Glue Mixture*. As the glue dries and flakes off, it carries the red spider off with it.

robber flies

From the illustrations it is easy to see that these Robber Flies are related to the house fly. Even though they look rather unkempt, they are all helpful predacious insects. More than that, they all are common everyday flies which one meets a dozen times during the summer season—that is, if one does not kill them off with insecticides.

The robber fly, pictured here in both male and female adult forms, is well known and is very often encountered in the home garden. It is an easy one to recognize, once we have made its acquaintance, by its suddenly-tapering abdomen, coming to a thin point at the end. This type often alights on the soil as well as on the plants, and it is often to be seen watching for another insect to seize and devour. There is an ironic but true-to-life picture in a popular insect guide showing this robber fly eating a lady bug!

Still another model of robber fly is a downy individual who looks at first glance like a bumble bee. This is no doubt the reason for its scientific name of *Bombomima.* This one gets special mention because it is a known predator of the Japanese beetles. One writer tells how Bombomima catches and kills Japanese beetles, holding them in its forelegs and eating until there is nothing left but the empty beetle shells.

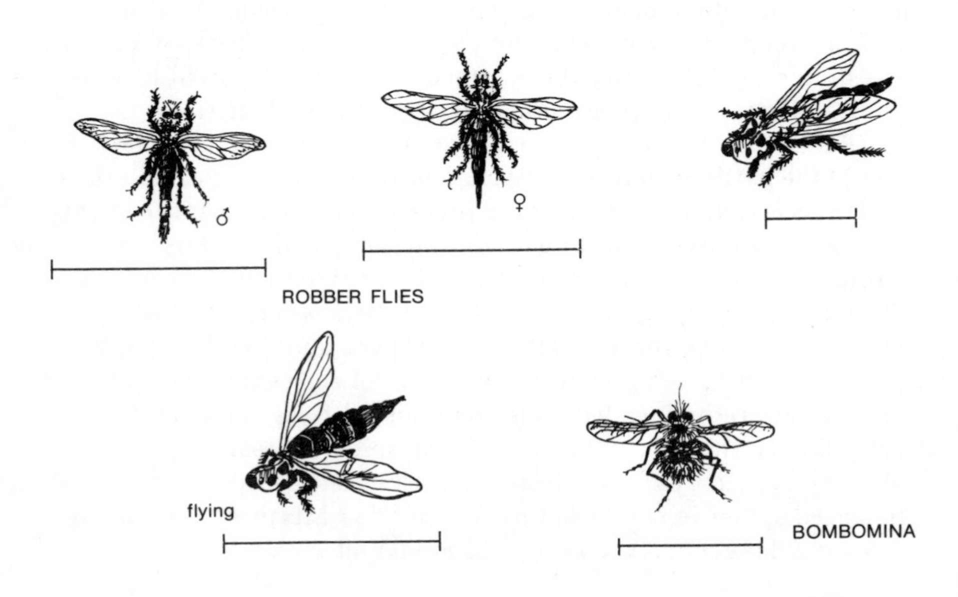

root maggot fly

There is a large family of flies which look very much like the familiar house fly. Each has a taste for a different garden vegetable. Cabbage, carrot, turnip, onion and seed corn—each has its own enthusiast. The adult female fly lays eggs on the roots. The maggots hatch out and live on the roots, thus weakening the plants. There is one simple remedy: wood ashes sprinkled liberally around the stems of the seedlings. If it rains and the wood ashes become soaked, replenish the supply with clean fluffy ashes, preferably fresh from the fireplace. We do not know exactly what happens, whether the soil structure is changed so the fly will not lay her eggs, or whether there is something in wood ashes which is a repellent. The simple fact is that all root maggots can be controlled by this common material.

Another method of control for cabbages is to place circles or squares of tar paper on the soil around each seedling. The scent of the tar repels the fly so she will not lay eggs on that seedling. Discs may be cut from tarred roofing paper. Slip stem into slit in disc so paper fits tight around the stem of the seedling.

ROOT MAGGOT FLY

TAR PAPER DISC

Still another way to control the root maggot is to dig up the soil in autumn and leave the pupae of the root maggots exposed to the frosts of winter and to the attacks of hungry birds.

It is possible to finish off carrot maggots for good, simply by dipping full grown carrots into hot water.

Loudon writes, "Carrots, when they come up, are apt to be attacked by insects, like the turnips: the most approved remedies for which are thick sowing, in order to afford both a supply for the insects and the crop, and *late sowing,* especially in light soils, thus permitting the grubs to attain their fly state before the seed comes up.

This is another general principle of insect control: *Plant any crop at a time when its particular pest is in an inactive stage.* Sow early when it is too cold for the pest to be abroad. Or sow late when they have already reached the fly stage. This applies to many vegetables. But one needs to be very familiar with each insect's habits and preferences in order to make this method of control work.

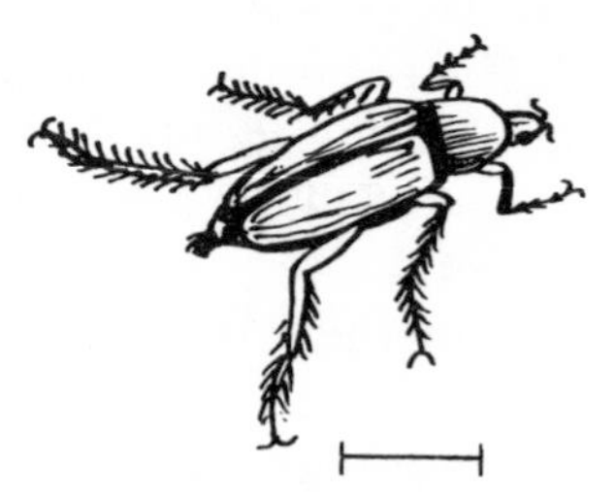

rose chafer

Here is a familiar but very unpopular little intruder plucked off our favorite rosebush. He bears a fine Latin name which means "long fingered" and one can see why he earned this title. He is also called Rose Beetle or Cock Chafer but under every name he acts badly, eating roses and grape blossoms—and we now suspect even the blooms on the elderberry bushes.

If the soil is plowed deeply and left all winter in clods, the young will be exposed to the cold and to the birds, but it is not very often practical to plow up a lawn. If is possible to sprinkle naphtha flakes or moth balls on the soil and to rake the soil to mix the naphtha thoroughly, if one must kill the grubs close to a particular rosebush. Remember, however, that one is also killing soil bacteria which help keep that rosebush in a state of health.

It is also possible to collect the beetles by hand several times a day, dropping them into a cup of kerosene, which kills them.

We had good success in driving them from one shrub at a time by Handpicking and Letting Insects Decompose*. We put a little water in a glass jar and filled the jar with rose beetles, letting them decay in the jar with the cover on. When there was a bad infestation of beetles in one shrub, we left the jar, with the cover off, underneath that shrub. The beetles found it distasteful and vacated that shrub. We didn't blame them! This is an absolutely harmless way to rid the whole garden of a long fingered pest of unbelievable persistence and tenacity. Our neighbors were very envious—but went on spraying with poisons.

scale insects, coccus and **mealybug**

Scale Insects look like tiny domes of various shapes and sizes as they cling tight to the twigs of the plants they infest. They have hardly any legs at all and they look like shells flattened on the tree twigs. The lady bug and her larvae are most efficient in keeping scale insects under control.

Scale on fruit trees may be controlled by applying Spirit of Turpentine* with a soft brush very lightly on the scale insect itself, so as not to injure the bark with the turpentine. This is parallel on a large operation to using a cotton swab dipped in Denatured Alcohol* to control mealy bug on house plants, (See Mealy Bugs).

For Coccus Scale (See Coccus-Codling Moth) on apple trees: wash tree trunks and branches carefully with soap and water* and a natural sponge, and always destroy *each scale as soon as you see it.*

Glue Solution* in a thin spray will harden and imprison the scale and render it powerless to do any further harm. Scalecide is a commercially-made dormant oil spray, available from B. G. Pratt Co. 206 21st Avenue, Paterson, N.J. 07503.

slugs and **snails**

Although neither the Slug nor the land Snail is an insect, the damage done by either one is comparable with the worst any insect can do. Lime or salt or soot will kill both of these pests on contact. However, if a slug is once sprinkled with salt, it gives off "a slimy exudation with which the creature is protected. If again and quickly

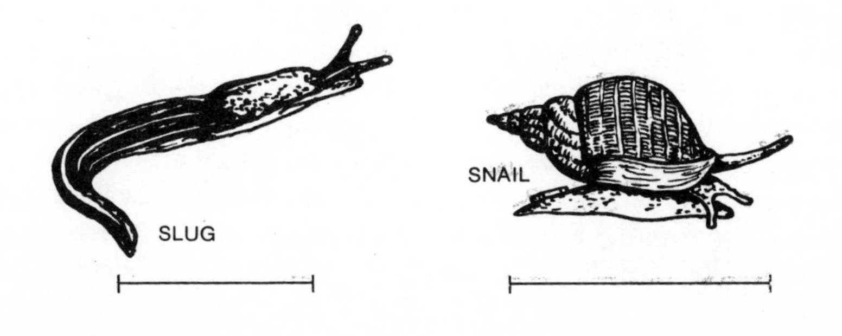

assailed in a similar manner, death is certain to follow. If one salting or liming does not answer, a second is likely to prove completely effectual."

We had a large area where plantain grew profusely, close by a vegetable garden. At dusk the slugs used to leave the garden to gather on the plantain leaves which they ate with pleasure. As it grew darker, they used to climb higher on the leaves until they were in complete view. It was not difficult, although it was not very pleasant, to sprinkle them with the kitchen salt shaker. When the table salt touched their slimy bodies the second time, as told above, they shrivelled and disappeared and we have never had trouble with slugs in that place again.

Slugs are attracted to various baits: sawdust or lettuce or cabbage leaves, or sliced turnips or potatoes, or grapefruit shells set on the ground like domes. Any of these baits should be placed at dusk and gathered in the morning when they should be buried or burned if there are no pigs to feed them to. A saucer of fresh beer sunk level with the ground works as bait to attract and kill slugs. A still better bait is very thin bread dough poured on the ground in a few places around the garden. The fermenting dough attracts the slugs irresistibly and results in their demise.

The bitter principle in an oak leaf mulch (see Mulch*) will repel both slugs and snails. Wormwood Tea* works on the same principle —the bitterness is exceedingly uncongenial for these slimy, tender bodied creatures (which can eat *so much* in a single night).

spiders

One of our most valuable insect controllers is the Spider, that insect-eating member of the family of invertebrates. All true insects have only six actual legs, but every spider has eight legs. All eight legs grow out of the actual body of the spider, while on an insect the legs emerge from the part called the thorax.

Spiders are best known for the webs they build—to the horror of the housewife! The purpose of the spider's web is to set a trap to catch insects. It is called a cobweb from a Latin word which means to seize, *capere*, and the spider in olden times was called *coppe*. There are hundreds of different spiders, each with its own habit of building a web—some in a hole in the ground, some on the grass,

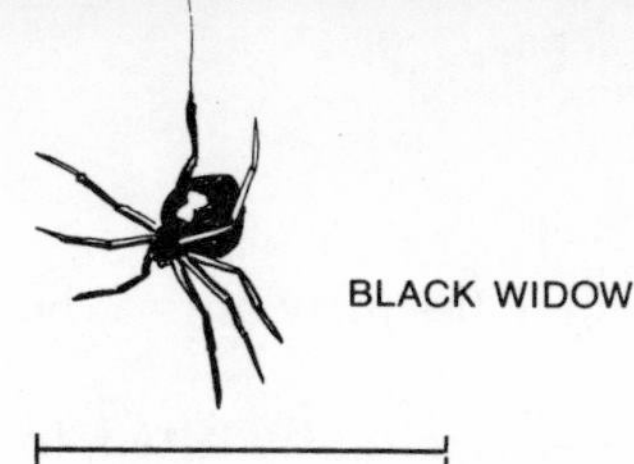

others on garden plants and some even in the trees. The spiders that live in the house earn their keep by catching and eating as many houseflies as come near. At the end of the winter one may find an undusted windowsill cluttered with moth wings below a spider's web. The spider spent the winter capturing moths and eating their bodies, but he did not care for the wings. He will also reject the wings of houseflies after eating the body.

Some spiders even will capture and kill grasshoppers two or three times their own size. A very absorbing spectacle is a battle between a wasp and a spider. Because they are so evenly matched, the outcome is unknown and the suspense remains high until one or the other wins.

The tarantula is poisonous and the black widow is somewhat poisonous if she bites a human. However, she rarely bites unless cornered. She is called a "widow" because she has eaten her mate! The illustration shows the characteristic hourglass marking on her smooth, shiny, jet black back. The hourglass marking is in red or bright orange. She is handsome, but wicked!

squash bug

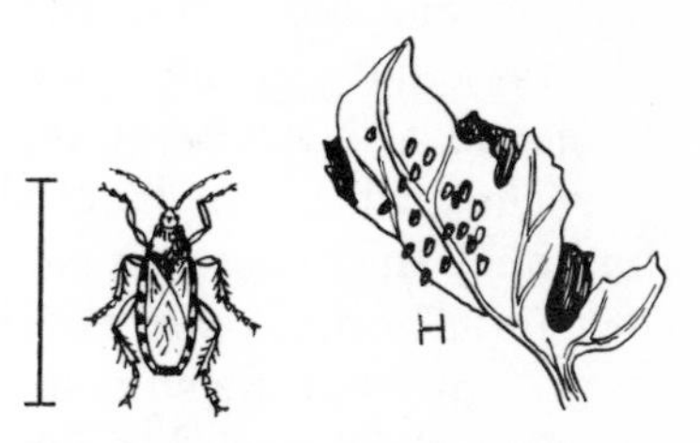

This illustration of the adult Squash Bug easily explains why it is also called one of the family of "shield bugs." If you should catch one of this large family in the garden or on squash, raspberry or other plants, you will also know why it is sometimes called a "stink bug." Of all the insects, this family is one of the few correctly called *bugs*, because the term bug refers to an insect which develops from egg to nymph and finally to an adult with only two wings and no hard wing covers. The clusters of eggs are golden yellow and encased in a hard shell. If eggs are located and crushed as soon as they appear, you will have no trouble with squash bugs for that season. If some escape notice and hatch, they will grow into repulsive gray nymphs with fat bodies and black legs. They suck the juice out of the squash plant.

Traps* may be made by laying thin flat boards, slightly tilted, in the garden rows. The squash bugs assemble beneath the boards and may then be easily crushed. Another control is Turpentine and Wood Ashes*. Another is Hydrated Lime and Wood Ashes* which may be applied from a sprinkling can. If the insects are sprinkled when they first appear, they are kept from multiplying beyond control.

Sow radish seed in the hills with squash or cucumber seed and they help to repel the striped cucumber beetle when the vine crops are very young. Nasturtiums sowed in squash hills will repel squash bugs.

squash vine borer

The adult of this species is a moth which is rarely seen. And if you should see it, you might not recognize it as a moth because its rear wings are transparent. Its color is an iridescent greenish-black, and those who know say it is very handsome. It lays its eggs on the stem of the squash vine—or other cucurbits—where the grub hatches and bores into the stem of the vine. When the grub finally fills the cavity in the hollow stem, the vine wilts because it can no longer sustain itself out beyond the borer.

The grub can be located by splitting the stem. Cut out the borer and bury the incision well under the soil so it will heal and will not dry out. The other parts of the vine will continue to grow.

To prevent the squash vine borer from settling on your vines in the first place, sprinkle camphor or black pepper around the roots of the growing plant to repel the moth. Be sure it is real camphor and not naphthalene or some other chemical substitute. The moths may also be repelled by Turpentine* fumes or by Aromatic Herbs*. The

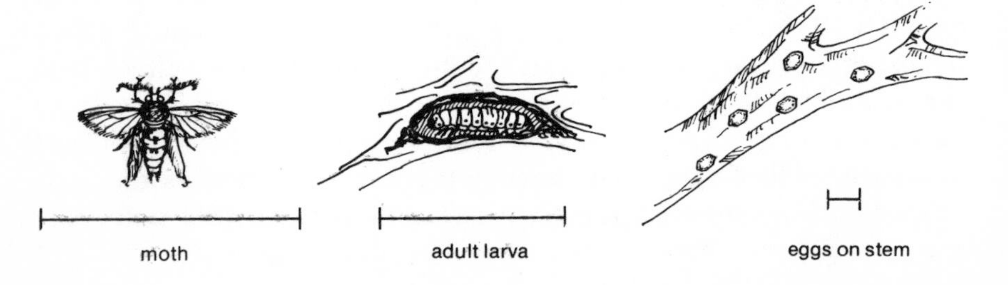

moths might also be attracted and trapped by an Electronic Insect Trap*.

The borer may be found on squash, pumpkin, cucumber, melon or cantaloupe vines, and the damage is recognized when the whole vine wilts from a certain point to the end of the vine. Do not confuse this serious wilt with the daily habit of the vine leaves to droop slightly in the hot afternoon sun to protect themselves from too much evaporation.

tarnished plant bug

This active little fellow is not listed in some of the larger federal entymological writings. We thought, therefore, that we would overlook him, lest this book become overcrowded with impractical details. However, the last celery plant we harvested in our garden sent us scurrying through the bug books to find out all about him.

In the first place, this is another true bug,—six legs, two antennae, two wings but no wing covers. It is very common and attacks all kinds of vegetables and fruits. Because it is so very active, and so shy of interference, it is seldom noticed until much damage has been done. It prefers to puncture the stalks near the joints. The nymphs are wingless and green, and they move very quickly when disturbed.

One of the State Agricultural College Bulletins gives instructions to dust this small villain with DDT, but the same bulletin warns not to use DDT on any edible crop! One of the old books recommends using soot for other kinds of flies on celery. Perhaps soot* would be effective for the tarnished plant bug. All indications point to our using soot next year in the celery patch.

Damage to celery in particular is recognizable by dense brown patches on the leaves, a stunted appearance of the leaf tips extending halfway down the stalk of the tenderer parts in the center of the celery stalks. During the growing season, in spite of the vigilance of two curious gardeners, there was absolutely no sign that this little bug was busily at work undermining the celery patch from within.

tent caterpillar

The Tent Caterpillar is easy to identify because the tiny caterpillars spin a web around themselves where they live in cozy comfort in early spring. If this web is wiped off the crotch of the tree with a cloth soaked in kerosene, it is the end. Do not try to burn out the nest because you will kill the branch too. If the caterpillars get out and grow up, they will travel long distances and will eat great quantities of leaves. One effectual way to control the caterpillar is to gather and destroy the egg cases during the fall and winter, when it is easy to see them, cutting off the twigs with pruning clippers.

Another way to reduce the population of tent caterpillars is to remove and destroy the tough white cocoons wherever they are found—under dry bark or in other crevices in the trees. Since the tent caterpillar is especially attracted to wild cherry trees (a trap crop*), this is a good sure place to locate them to destroy egg cases or cocoons. Also see Baits*. See also Caterpillars.

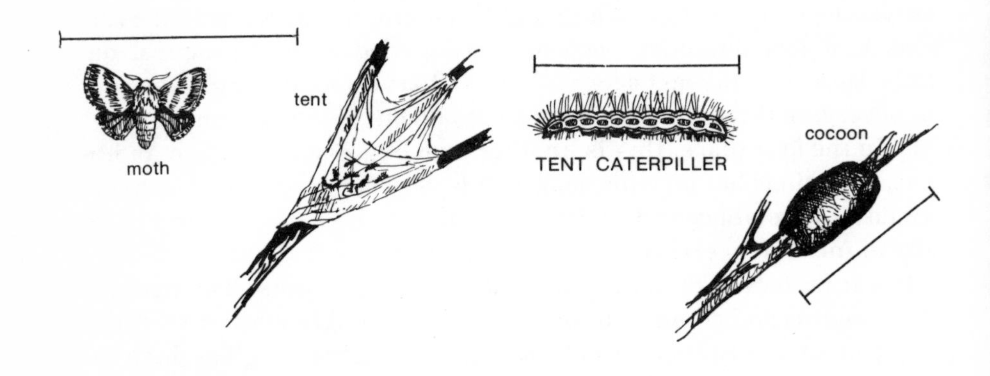

termites

Although the Termite is not actually a garden pest, it sometimes is found in the vicinity of gardens, and should be included here if only for identification. The termite looks like an ant except that it does not a "wasp-waist" like the ant. Termites have two pairs of wings which are all of equal length, whereas winged ants have long front wings and shorter hind wings. Their actual size is less than one half inch long.

Termites must have contact with the soil even though they are living in wood. For this purpose they make small tunnels of clay or earth which stretch between their dwellings and their food supplies. A government bulletin states that termites may be controlled simply by scraping away the earthen tubes in order to break this connection between nest and food. The same bulletin also says that termites do not cause a sudden and devastating destruction. . . . It usually takes them a number of years to cause extensive damage to a building.

Sometimes the home gardener who has left wooden flats too long on the ground will find the bottom of the flat invaded with small white grubs where the wood is in contact with the soil. If flats are stacked where they have plenty of air, lifted off the ground by a brick or two, there will be no danger of infestation by termites. This also will help preserve the wood so the bottom will not fall out some day while you are carrying a flat full of tenderly nurtured seedlings.

thrips

Two tiny specks would indicate the true size of the Thrips. The enlarged drawing is magnified nine times! Some flowers like the white daisy may be covered with these slender, quickly moving little rascals who seem to turn and twist and slither sideways as well as forward. On the wild flowers they make little difference, but when they attack our onion patch or other plants we are trying to raise, they are not exactly welcome. Thrips can decimate an onion bed in short order because they are so small that they are unnoticed until the damage has gone too far. However, they do leave a black deposit of tiny specks which can be seen, and the debilitated condition of the plants shows that something is causing trouble.

It is most important to keep roots growing vigorously and this can be accomplished by the use of Bio-Dynamic Preparation 500 which might even help prevent an invasion of thrips in the first place.

Once thrips have a foothold, they can be brought under control by a mulch of tobacco dust* or by a strong fumigation of tobacco smoke* two or three times when the wind is calm. Another method somewhat easier to control is Tobacco Water and soft soap solution*.

Since *some* kinds of thrips are helpful on some plants, it would be better not to kill them unless they are actually causing damage, but it takes much patient study to know whether they are friends or foes.

tomato hornworm

Practically everyone who has ever raised a garden has shuddered with horror at the sight of the Tomato Hornworm because of its sheer size and voraciousness. There is no better way to keep ahead of this "beast" than to steel oneself to the experience and hand pick. It is possible to place the caterpillars underneath a large stone and then, with determination, step upon the stone!

It is also possible to attract the moths to a lighted trap at night and thus to prevent their multiplying in the home garden. See Electronic Insect Trap*.

Someone complains that the tomato hornworm is almost impossible to find in the garden. It does take careful inspection, but once

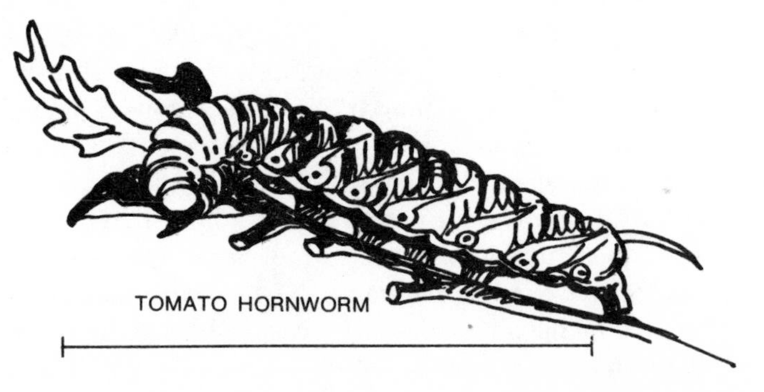

you know the secret, it is very easy to find one: look first on the ground underneath the tomato plant *when* you first come out early in the morning. If you have an active tomato worm—and who has an inactive one?—he has been eating voraciously all night and the earth directly beneath him will be littered with dark green droppings. Look at the plant above, and after a few minutes concentration, your eyes will tell which is the caterpillar and which is a curled up leaf. This worm is one of Nature's best examples of protective coloring, and that is why you have to search so hard to find him.

The hornworm eats all the tomato plant he can hold and then one day he becomes very quiet. In a little while he has turned into a pupa, and we sometimes find these pupae in the ground or tucked away in a safe place to spend the winter. They can be identified by the free tongue case.

If nothing happens to the pupa during the winter, when the conditions are right in the warm spring weather the moth comes out of the chrysalis. This is one of our larger moths, although its coloration is rather sombre and dull.

We have explained in detail these three stages of the development of this moth. The butterfly family follows the same pattern: egg, larva or worm or caterpillar; pupa or chrysalis or cocoon; and finally the butterfly or moth, which is the adult and perfect state of the insect. If the particular insect does damage in one state, he may be beneficial in another. If he cannot be captured or controlled in one state, he may be subject to predators in another state. An intimate knowledge of the various life cycles of all insects is necessary for home gardeners, so that they may work with the good bugs and against the bad ones to the best of their knowledge and skill.

TOMATO HORNWORM moth adult

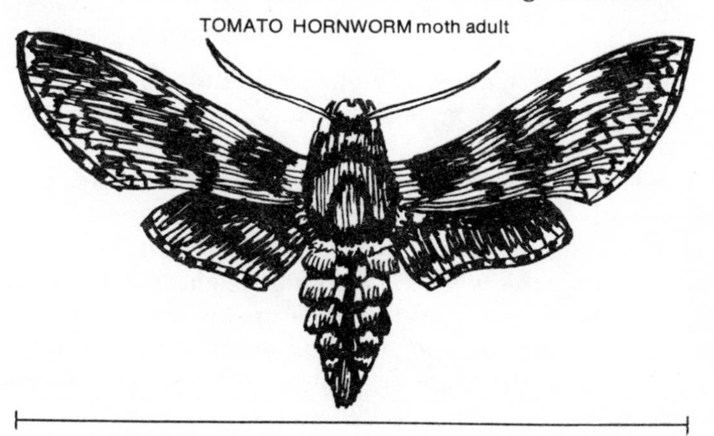

tree hoppers or **leaf hoppers**

Even the serious insect book says that the Tree Hoppers are the "Brownies" of the insect world. One often meets them on leaves and grasses in midsummer.

And one can only look with incredulity at the ridiculous shapes of these comical and out-of-this-world creatures.

Their record of behaviour is not too good. They sometimes eat more than their share of tree leaves, but so far they have made up for the harm by their charming and grotesque appearances!

weevils

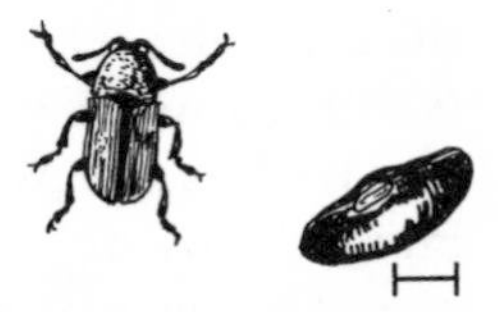

The Bean Weevil originates from the egg of a tiny beetle which the female laid in a bean blossom. It is, therefore, advisable to plant beans at a time so the blossoming stage will not coincide with the flying stage of the adult bean weevil. If the beans are sown as early as possible in the spring—noting that beans are tender and must be sown after frost is past—they may escape the visits of the bean weevil moth. Similarly, late sowings of beans may be timed in each locality to be too late for the bean weevil moth's flights. To determine the exact days when there is danger from the moth would take some trapping and some study of local habits of the insect. Your County Agent may have helpful information here if this is an insect of bad reputation.

It used to be a regular practice when beans or peas are dried and stored for winter use, to move them every week or two. Dr. Ehrenfried Pfeiffer tells of grain stored at his farm in Holland where one of the miller's chores every week is to go through the tremendous storage rooms and turn every bag of grain upside down. The next week he goes through and turns every one right side up again. This continual upsetting of the grain is also upsetting to any insects which might otherwise take up residence in the stored grain.

Beans or peas stored on a small scale in a bin or keg, can often be stirred with the hands to be kept free of weevil damage. Wormwood tea* sprinkled where grain is stored will repel grain weevils. See "A Prime Method of Insect Control" also, at front of the book.

wheelbug

Any gardener might be confronted right in his own backyard, as we once were, by an incredibly homely gray insect of rather large size, making himself quite at home in the shrubbery and on the sunflower plants. When viewed head on, it looked ferocious enough to devour one. In profile it was carrying an enormous projection on its back which looked like a wheel with cogs all around the edge.

We looked him up in the book. The wheel was attached and this was a real live Wheel Bug.

This is another of the predators which is to be treated kindly, because he is one of the good ones. He eats other less hard-shelled insects. He has a certain favorite post where he stands guard, and since then we have often watched him lying in wait for unsuspecting insects which might come his way.

The only warning to human beings is not to tamper with him: his temper is irrascible and the bug books tell us that he can very swiftly turn and administer a mean bite.

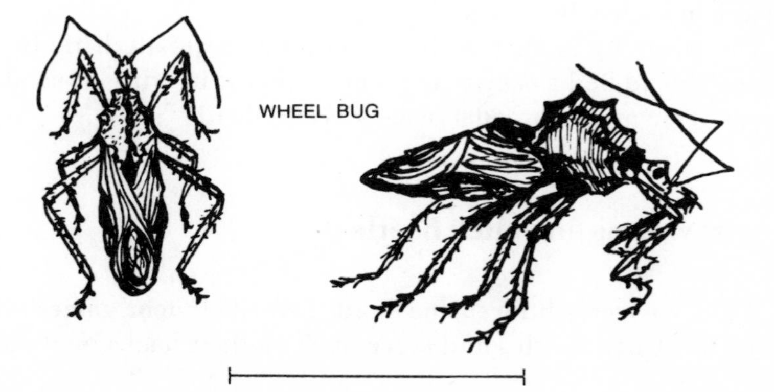

white fly

White Fly is a very small aphis-like insect which looks somewhat like a very tiny moth. It is often found in the greenhouse, where it is especially fond of peppermint and other edible herbs. Because of the dangers involved in spraying edible plants with poisonous substances, experiments are being made to find repellents for the insect. It has been established that a few plantings of rhubarb in the greenhouse will repel the white flies.

Out of doors the white fly is harder to control, and one might be forced to use tobacco dust* or to spray with a nicotine and soap solution* or, in very bad cases, to use Kerosene Emulsion*. Be sure to wash off afterwards with clean water. The treatment may have to be repeated several times because the nymphs are difficult to reach on the underside of the leaves.

The white fly treatment does not, alas, permanently eradicate, but it does cope for quite a while; then has to be repeated.

"I use one fluid ounce tincture of green soap and two tablespoons Ryania powder to one quart cold water. Mix thoroughly and pressure-spray on all surfaces of infested plants.

"We've had no serious problem after using this spray twice at one-week intervals*."

It is good news to hear that people are working on non-poisonous repellents for white fly—and for other insects. Botanists, entymologists and bio-chemists are making up teams in some of our universities to study the interrelationships of their various departments. It is to be hoped that more money will be appropriated for research in such enterprises.

The white fly is such a tiny, charming, exquisite little bit that humans should find a way to repel or to outwit it, without sending a barrage of chemicals broadside to exterminate it!

wireworms and **click beetles**

When we were children one of our favorite indoor games was to catch a few Click Beetles and place them on their backs on a smooth surface—often in the bathtub with no water in it—to watch them leap high in the air with a sharp little click. Ninety-nine times out of one hundred they landed on their feet and tried to hurry away, while we scrambled after them and made them jump again and again. One could hold the beetle tight in one's fingers and feel the sharp snap it made as it tried to get away. There is a very strong hinge between the thorax and the abdomen which causes the snap. They are sometimes called Snap Bugs as well as click beetles.

Little did we know in those carefree days that this amusing playmate was the adult state of an insect which causes much trouble in all the branches of its family.

It is the hard, shiny Wireworm or Brown Wireworm, which one finds in the topsoil of the home garden. In the potato patch the wireworm is not welcome because it burrows into the potatoes and leaves tunnels. Actually its taste for potatoes may be its undoing, because the quickest way to rid a garden of wireworms is to put out pieces of potato as bait*. The important thing is to examine the bait pieces every week to remove the wireworms.

Another member of this same amusing family is the Eyed-Elator (from the Greek "to drive"). The simulated eyes on his back make him look like a ferocious eyed monster and protect him from birds which might otherwise eat him. His true eyes are where they should be, in front. His larvae eat other soft-bodied insects. He also *snaps*.

woolly aphis (also see aphis)

When an entymologist writes about the Woolly Aphis, he tends to be terse: "The Woolly apple aphid is a reddish or purplish aphid which is covered with white powder or fibres and occurs around wounds or water sprouts or roots of apple trees. It has a complicated life cycle involving generally both elms and apple trees and causes considerable damage to the roots of old apple trees." Compare this with the following:

"I know of no more diverting occupation than watching a colony of aphids through a lens. These insects are the most helpless and amiable little ninnies in the whole insect world; and they look the part, probably because their eyes, so large and wide apart, seem so innocent and wondering. The usual color of aphids is green; but there are many species which are otherwise colored. . . . In looking along an infested leafstalk, we see them in all stages and positions. One may have thrust its beak to the hilt in a plant stem, and be so satisfied and absorbed in sucking the juice that its hind feet are lifted high in the air, and its antennae curved backward, making all together a gesture which seems an adequate expression of bliss. . . . It is comical to see a row of them sucking a plant stem 'for dear life', the heads all in the same direction and they packed in and around each other as if there were no other plants in the world to give them room, the little ones wedged in between the big ones, until sometimes some of them are obliged to rest their hind legs on the antennae of the neighbors next behind. . . .

"A German scientist discovered that a plant louse [aphis] smeared the eyes and jaws of its enemy, the aphis lion, with wax which dried as soon as applied. In action it was something like throwing a basin of paste at the head of the attacking party; the aphis lion thus treated was obliged to stop and clean itself before it could go on with its hunt, and the aphid walked off in safety. The aphids surely need this protection because they have two fierce enemies, the larvae of the aphis lion and the larvae of the ladybirds. They are also the victims of parasitic insects; a tiny four-winged "fly" [perhaps a Braconid Wasp] lays an egg within an aphid; the larva hatching from it feeds upon the inner portions of the aphid, causing it to swell as if afflicted with dropsy. Later the aphid dies, and the interloper, with malicious impertinence, cuts a neat circular door in the poor aphid's skeleton skin and issues from it a full-fledged insect. . . .

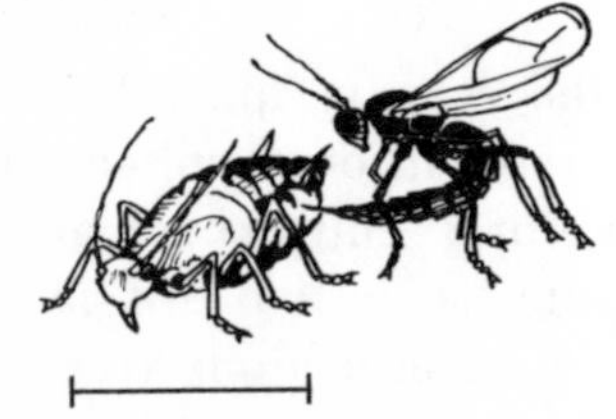

WOOLLY APHIS—parasite

parasite emerging from WOOLLY APHIS

"Plant lice vary in their habits. Some live in the ground on the roots of plants and are very destructive; but the greater number of species live on the foliage of plants and are very fond of the young tender leaves and thus do great damage. Some aphids have their bodies covered with white powder or with tiny fringes, which give them the appearance of being covered with cotton; these are called Woolly Aphis." Comstock, A.B., HANDBOOK OF NATURE STUDY, Ithaca, New York.)

The Woolly Aphis was so bad and so prevalent in England in the last century that it was named the American Blight. Sutton says, "We have seen old trees restored to youth by scrubbing them with dandy brushes dipped in hot brine. Thus must be done during winter, or before the trees come into leaf in spring, and mats must be spread to catch the splashes or they will kill the grass under the trees. A careful pruning should accompany the washing and the prunings should be burnt. In the course of the summer the woolly pest will appear again, and should be extirpated by carefully washing the patches with methylated spirit*. Fir Tree Oil Insecticide is a sure remedy for Woolly Aphis; and pure water will go some way towards cleansing the trees if well brushed into the wounds this destructive insect produces in the bark of the trees. A good paint for Apple Trees may be made with Gishurst Compound at the rate of 8 ounces to the gallon of water, with a little fine clay added to render it adhesive." Sutton, THE CULTURE OF VEGETABLES AND FLOWERS, London, 1890.)

Does anyone know anything about Fir Tree Oil Insecticide or Gishurst Compound?

other pests

BIRDS—The principle for control of birds that makes the scarecrow effective is that if birds believe there are human beings nearby to threaten, they will be scared away—provided the scarecrow is realistic enough and moves often enough. Within a surprisingly short time, however, the birds find out whether or not the scarecrow is really a threat to them. Working on this same principle, it may be possible to frighten the birds with white strings on which there are tied many white rags every 10 to 15 feet. These white rags should be put up before the birds have discovered the crop that is being protected. If birds find out first that there is food to be eaten, no scarecrow will ever deter them.

There is available a three-dimensional model of an owl made of plastic, 12 inches tall, hung by a swivel from the top of its head. This is said to look so much like an owl, and to behave so realistically, that the birds are frightened away and do not come back! Available from Farmer Seed and Nursery Co., Faribault, Minn. 55021.

Another protection product is plastic netting which comes in a number of sizes and colors to keep out birds as well as smaller pests. This is manufactured by a firm which sympathizes with the home gardener's desire to protect his crops from the birds. Netting may be purchased from most seed houses and from agricultural supply companies like Agway or from Animal Repellents, P.O. Box 168, Griffin, Georgia 30223.

DEER—High wires strung around the garden with white rags fluttering from them, too high for the deer to jump over, will protect garden crops. We have been told by people who know that human urine will also repel the deer. Dried blood works but must be replaced after rain.

FOX—The fox stalks his prey as a cat does. He does not look at the fences because he has his eye on the animal he is stalking. An electric fence with 5 strands, 3 inches apart, with the bottom strand 3 inches above the ground has been effective in excluding foxes from home gardens. There is no need for a woven fence if you use electric fencing.

MOLES—A mole repeller windmill advertised in current magazines whirls on a pole driven into the ground. This makes a drumming noise and vibration in the ground which will drive away the moles. Two such mills are supposed to clear the moles from an average city lot. Available from Mother's General Truck Store, Box 506, Flat Rock, NC 28731.

Of course traps may be used: "The true mode of getting rid of moles and one most readily put into execution is to dig up their nests in spring. The heaps of earth over these nests are easily known from common mole-heaps by their size," says Loudon. The most effective mole-exterminator is the family cat!!

MICE—Nut tree leaves spread over stored fruits will protect the fruit from mice. Sassafras bark scattered in stored fruit will repel mice. Camphor gum also will keep mice away. Protect seeds by mixing with them small pieces of camphor gum. The plant Spurge (*Euphorbia lathyrus*) repels mice. Catnip is supposed to repel rats and mice, perhaps because it attracts cats which are their strongest enemy. Mice are repelled by *dog fennel* scattered in the granary. "Mice may be kept under by the different domestic traps, or by the gardener's or four-figure trap (see TRAPS below) or by an earthen vessel with a narrow mouth, and bellied out within, sunk in the earth and a few leaves or straws placed over it, as is common about Paris. (The mouse falls in and cannot climb out because of the shape of the jar.) But two or three cats kept in a garden, are the most effectual destroyers of mice." (1824). Automatic mouse traps are available from seed catalogs and supply houses.

RABBITS IN THE GARDEN—A row or two of soy beans planted along the edge of the garden will keep the rabbits busy until green beans have reached maturity and are no longer attractive to them. Plant green beans in hills rather than in rows and as soon as the plants appear, or even before, surround each plant with a coil of chicken wire 12 inches high. This year I learned from a game warden of a method so simple and practical as to make one wonder why it has not been generally adopted long ago. Dissolve two tablespoons of Epsom salts in one quart water. Sprinkle on bean plants. It will not injure the plants or the human consumer. The rabbits, however, detest it. A fresh application is necessary following a heavy rain.

Scatter moth balls to repel rabbits. Several people have witnessed the value of this practice: planting a row of onion sets all around the home garden keeps the rabbits out. If you don't like onions, perhaps your neighbor does! Another rabbit control is to soak a rope in Creosote and stretch it around the garden.

"Where the hare is injurious by barking trees, smearing the stem with cow-dung, ordure (dung), tar, or coal oil (kerosene) will deter them," says Loudon. Rabbit manure, softened in water and painted on fruit tree trunks is reported to protect those trees from girdling by rabbits.

RACCOONS—This is one of the cleverest of the animals. We used to be resigned to losing corn to the raccoons unless we were willing to watch with a shotgun as they arrived. Recently we have met two different farmers who solved the raccoon problem with a small radio in a plastic bag set up in the cornfield and tuned to an all-night radio broadcast. They report that rock music is most effective as a raccoon repellent! The raccoon will let the corn alone if each corn ear is smeared with lard and red pepper on the outside of the husks.

RATS—Instead of enumerating the modern ways of outwitting this rather wary nuisance, we would like to quote Loudon who gives a little insight into the principles behind the use of the rat trap: "The garden rat trap should generally be a box or enticing engine of some sort, rather than a toothed iron trap; because unless there is a great scarcity of food, which is seldom the case in the field rat, it will not be allured by the bait of the former; whereas a trap may be so disguised by straw, or moss, or leaves, and so scented by oil of anise, as to be resorted to or at least not recognized by the rats till they are taken."

WOODCHUCKS—Even though the woodchuck is a burrower, he will not dig under a fence to get garden produce if there is plenty of wild food for him on the outside. Nor is he a climber—up fences, anyway. A six-to-twelve inch fence is high enough to frustrate the woodchuck's attempts to eat early peas and lettuce. Later in the season, when there is plenty of food anywhere, the garden is safe without any fence. However it helps to have a dog and a cat to keep the woodchucks moving—as we saw last winter when our cat forced a woodchuck to climb about six feet up a tree trunk. We were so surprised we could hardly believe it—and in broad daylight. All the above pests may be caught alive and unharmed in the Havehart Trap, available from seed and supply stores.

And finally—"Of the Kinds of Vermin most injurious to Gardens;" "*the human enemies of gardens* are such as break in secretly to steal clandestinely, to injure or destroy; or under the guise of regular operators, pilfer and otherwise act as enemies to the garden and its proprietor. The operations for deterring and detecting thieves are; watching by men, by dogs, by peacocks and turkeys allowed to sit on high trees, and by ducks. The dog is most effectual, but peacocks and ducks are known to scream or cry on the approach of strangers in the night time. As neither of these birds scratch the earth, they are in some kinds of gardens, especially nurseries, more useful in picking up insects than they are injurious. Man-traps, spring-guns and alarums are also set to detect and deter, and the notices of these dreadful instruments as well as the fear of the law, have considerable influence."

recipes and formulas

ANISE OIL

Fishermen and hunters who are out-of-doors in the early spring when gnats and "no-see-ums" are rampant, might find some relief in the USDA suggestion to wash their clothing in soap suds to which Anise Oil has been added. It is reputed to repel Houseflies and their relatives, the Green Bottle Flies and Black Blow Flies.

We also have a report of another use for Anise. Buns were treated with aniseseed and were fed to police dogs by the people being hunted. The dogs licked the anise off the buns and became very affectionate, and no one suffered any dog bites. Anise acts as a love potion, according to the report!

AROMATIC HERBS

The list of aromatic herbs is almost endless. There is no limit to the combinations which may be employed to repel moths: wormwood, santolina, mint—either fresh or dried—lavendar, rosemary, sage and southernwood, and most of the rest of the herb garden inhabitants. One who has any of these herbs in the garden can experiment by drying them as fast as they come along during spring and summer. Keep them stored in tins away from light, heat and dampness. Label them well, because you will not recognize them in their dried condition. It is simple and exciting to mix different combinations to try as insect repellents at any time during the year. The above list of moth repellents will work indoors in stored clothing as well as outdoors in the home garden.

Tomatoes will keep the cabbage butterfly from attacking members of the brassica family but it is better to use a Tomato Plant Brew (see below) than to plant tomatoes near the brassicas. They are not good companion plants.

Sowing leek seed with carrots repels the carrot root maggot.

Lettuce will protect radish from the flea beetle when rows are sown in this pattern: lettuce, radish, kohlrabi, radish, lettuce.

Artemisia absinthium repels fleas, corn and wheat borer.

Artemesia vulgaris attracts flies.

Artemesia contra provides santonin, a remedy for intestinal worms.

Tansy also belongs to this family; keeps flies off fresh meat. When dried it will repel fleas *(Bio-Dynamics Quarterly.)*

Nasturtium repels squash bugs and also woolly aphis on apple trees.... *(Bio-Dynamics Quarterly.)*

Bacillus Thuringiensis, Thoracide. International Mineral and Chemical Corp., Crop Aid Products, Dept. 5401, Old Orchard Rd., Skokie, Ill. 60076

BAITS

Fruit Tree Moths. Tie quarter inch mesh (to exclude honey bees) over the top of preserve jars. An inch or two of brown sugar and molasses in the bottom of the jar will act as bait for any fruit tree moths that are passing by, especially if the bait is flavored with special aromas, for instance, sassafras for codling moth. Be sure to hang the jars in the trees early in the season before the trees begin to come into bloom. Other flavors which will attract—to be added to the brown sugar and molasses—are Oil of Mace and Pine Tar Oil.

Bait for Gypsy Moth. A federal entomologist has made a study of a gypsy moth bait called Gyptol or Dispalure made from the last two segments of the female gypsy moth. This bait is used in a trap to attract the male gypsy moths which are detained and later destroyed. This substance was extracted from female moths and now has been made synthetically. However the latest expressed opinion of the government entymologist is that this is not a large scale method to control gypsy moth.

Those of us who have worked with natural substances for a long time have noticed that oftentimes the "real thing" *works* because it has a dynamic living relationship to the changing forces and substances in its living environment, whereas its synthetic counterpart fails. Perhaps if the entymologists would continue to use the real sex attractant from the female moth, it would in the long run prove to be a practical way to invite all the male gypsy moths and lead them to their destruction, thus controlling the gypsy moth menace.

Bait for Houseflies. One successful housefly bait is stale beer added to other fermenting sugar flavors. The fly goes in search of this attractive food, enters a wire screen trap and then can be disposed of by burning.

Another bait: 1 part blackstrap molasses, 3 parts water, milk, or fruit juice.

Another bait popular with houseflies: any kind of decaying meat or fish placed in a small dish inside the wire screen of the fly trap.

Bait for Japanese Beetle. Bright yellow beetle traps may be baited with a commercially prepared bait. Japanese beetles may be attracted to white geraniums which they like, but pay little attention to red or to pink geraniums. Collect the beetles daily from the bait plants and destroy them with hot water or by burning. See also Doom and Electronic Insect Traps below, and Japanese Beetle.

Bait for Slug and Snail. Invert grapefruit skins on the ground where slugs are prevalent. Or spread lettuce or cabbage leaves every evening to lure slugs and snails. Slices of turnip or potato will also attract them. It is important to remove each of these baits while the creatures are still feeding. If you wait until the middle of the day, they will have returned to their nests and the baits will be useless. (See also page 78).

Bait for Wireworm. This one is easy because the wire worm is primarily a pest in the potato patch. Put pieces of potato—old ones—where the wireworm can find them, and they will let the growing potatoes alone. Collect the potato pieces every few days and dispose of the wireworms, and the area will be cleared of the pest before the potato crop is full grown. Or fill tin cans with potato and carrot peelings, after punching holes in the sides of the cans. Bury cans in the garden near potatoes and carrots. Wireworms will crawl into the cans for the bait. Empty cans each week and refill with bait.

BANTAM HENS

A few bantam hens provide not only an ornament to any garden but a lively and vigilant crew of animated insecticides. With the customary visual keenness of birds, bantams will spot an almost invisible insect and will nab it that instant. They are independent, requiring only the simplest housing and a minimum of feed. In return they will spend hours every day policing the garden and grounds looking for insects. They are extremely fond of strawberries and green tomatoes which should be fenced. Otherwise we found them diligent and sociable helpers. They multiply very rapidly (with a trio to begin with, we soon had two or three dozen!) and are a source of both eggs and good meat for the family which enjoys home-grown poultry.

BARRIER BANDS

Barrier bands on fruit trees and on other kinds of trees are put on to prevent crawling insects from climbing up. For instance, the wingless female of the cankerworm must climb up into the tree to lay her eggs. If she is prevented by a barrier band of some sort, the tree is protected from the progeny of that insect.

Barrier bands of cotton batting are also effective against caterpillars of the tussock moth, gypsy moth and cankerworm adult female. Bind a wide band of cotton batting around the trunk of the tree with a string. Tie the string securely around the trunk to hold the band near the middle. Turn down the upper part of the cotton batting, down over the string. This provides a barrier and trap which will protect the tree against any of these insects as they try to climb up the trunk into the tree branches.

Barrier bands of corrugated paper are also effective against caterpillars of the codling moth. Tie four thicknesses of corrugated paper around the tree with a thin wire, early in the summer. The grubs of the codling moth will crawl into the corrugated paper to hibernate. Burn the papers after frost in the fall to destroy the grubs.

Barrier bands of grease. The grease band is a ring of sticky material around the trunk of apple, plum and cherry trees to prevent the wingless insects from climbing up the trunk. The sticky material is available from all seed catalogs under a name like Tanglefoot Tree Paste, available from The Tanglefoot Company, 314 Straight Ave. SW, Grand Rapids, Mich. 49500 or seed houses. A recipe follows below.

We have often found it better to place a strip of paper or cloth around the trunk before smearing on the sticky material, which sometimes injures the bark. The most important point about these sticky bands is to get them on to the trunks by the end of September at the very latest. Otherwise the wingless adults of the codling moth and other fruit tree pests will calmly climb up the trees to lay their eggs in the tree tops where they will hatch the following spring. Then your orchard control program is delayed by one whole year.

Barrier bands of hay. One very inexpensive and practical method of trapping apple-blossom weevils, sawflies and codling moth caterpillars is to wrap strips of burlap around tree trunks in such a way as to hold a fair amount of hay. The hay band should be made ready in the spring right after the blossoming season so the insects can take up residence in early summer. Early in the fall the hay bands can be removed with all their inhabitants and burned.

Tanglefoot Tree Band Recipe:
- 5 pounds of resin
- 3 pints of castor oil
- Add 3 pints Venice Turpentine to this.
- Add printer's ink or very heavy, sticky oil.
- Apply this sticky mixture on a band of heavy paper 6 to 8 inches wide.

Tack or tie this on the trunk. Fill all roughness underneath the band with cotton to prevent insects crawling up underneath. Watch the bands and renew them when the stickiness wears off—or if they catch so many dead insects that the live ones are using them for a bridge to climb up anyway.

Barrier of wire netting. To trap Spring Canker and Fall Canker Worm bind a 14-inch wide strip of wire mosquito screen around the trunk of any fruit tree that is threatened. Tack the screen tightly around the trunk at the top of the strip, but let it stand out from the bark an inch or more at the bottom. You may have to support it on nails to keep it out. The wingless female of the cankerworm has to creep up the trunk to lay her eggs. With this wire "trap" she will be caught, and it is simple to crush her—and hundreds just like her—by pressing the wire against the trunk. They should be crushed at night. Watch to see any eggs the females may lay below the barriers on the tree trunks. For Spring Canker Worm apply the wire screen late in February or during March. For Fall Canker Worm put up screen bands in October and keep watch over them until the ground is well frozen in the fall.

BIO-DYNAMIC PREPARATION 508 EQUISETUM

Equisetum may be sprayed repeatedly to overcome fungus. Boil for 20 minutes: half a cup of Equisetum weed in one quart of water. Combine with Stinging Nettle Water (see below) to combat red spider mite and aphis, especially black fly. Fermented Equisetum Tea has been used, to which soft soap was added. Equisetum Tea has also been made up with 2% Waterglass solution. A 5% solution of waterglass helps trees against lice and fungus.

BLACK PEPPER

If Black Pepper is sprinkled around squash vines it will help to repel the squash vine borer.

CAMPHOR

Camphor is a whitish, crystalline substance made from

the gum of a tree from Asia. It has a very strong, pungent odor, strongly disliked by many insects, especially moths. It may be diluted in alcohol for finer distribution. Certain herbs in the garden—like sage, tansy, feverfew and the artemesia family—contain some camphor. Perhaps this is why they repel moths!

CHAMOMILE TEA

Make tea of Chamomile (or Camomile) flowers, as you would make it for a pleasant afternoon beverage, a heaping teaspoonful of flowers to one cup of boiling water. Let it steep in a teapot for a few minutes and strain. It is very enjoyable to drink! It is also a preventive of dampoff if sprayed on seed flats and flowerpots. Sprayed on cucumber seedlings, it will prevent mildew.

CHIVES TEA

Take a handful of chives cut fresh in the early morning. Put them in a teapot and pour on boiling water. Let the tea steep 15 minutes or so. Spray for mildew on gooseberries, cucumbers and other plants. The principle involved here is that chives are not attacked by any kind of mildew or fungus. Perhaps they have some quality which repels mildew which may help some other less resistant plants.

CLAY

Clay has a special quality of its own: its power to cover and to stick to anything it touches. Clay used to be used alone, in rather fine dilutions, and sprayed as an insecticide. If fine clay solution is sprayed over aphis, their tender bodies are rendered lifeless. Clay is also an important ingredient of the Bio-Dynamic Tree Spray because of its healing and sticking quality.

COMPOST WATER

Any plant which needs a special shot in the arm is given a few handfuls of good compost stirred into rainwater. Such treatment may help a plant's general condition when weakened by insects or by adversities in the weather. This is particularly effective for houseplants in winter.

CORIANDER OIL

The USDA Yearbook INSECTS reports that an emulsion spray containing a 2 percent addition of Oil of Coriander will kill the red spider mite.

CREOSOTE

Creosote has such a strong smell that it can be used to repel some insects. It is used in large-scale farming to soak paper to repel large invasions of insects. A bulletin on Termite Control recommends 1 part Creosote in 2 parts kerosene by volume to kill *termites* in the soil and to protect wooden buildings from their return and destruction. Creosote will burn plants if it touches them, however. Creosote rope may be purchased from most seed houses.

Within the memory of some of us, the "Bug Men" of each

New England town used to go with a small bucket of creosote and a long bamboo pole with a swab at the end. Wherever they saw egg masses of gypsy moth or tent caterpillar, they touched the eggs with the swab, which killed the eggs, so they never hatched. Tent caterpillars may be wiped out of tree crotches with a rag dipped in creosote. Egg-bearing cases like the bagworm may be dropped into creosote to finish them off.

DALMATIAN POWDER, see Pyrethum

DENATURED ALCOHOL

Denatured alcohol (called in England methylated spirit) is a deadly poison to every species of insect. Diluted with water it is effective. Used with caution it is inexpensive and effectual, and harmless to human beings. Dissolve some camphor in the alcohol and it will be still more effective. Do not let it touch plant foliage in its pure state as it will burn the leaves. Apply with cotton swab directly to insects. It will kill mealy bugs when nothing else will touch them.

DERRIS

Derris is the material out of which Rotenone is made. Rotenone is both a stomach and a contact poison to insects, but is non-toxic to humans or to wild or domesticated animals. Rotenone may be used on vegetables to be used for food, but it kills good bugs just as fast as it kills the bad ones at which it is aimed. Rotenone is available from seed houses or hardware stores.

DIATOMACEOUS EARTH, see Kieselguhr

DIPEL

Dipel is an organic control for insects in the caterpillar stage. Galt Research, Inc., RR # 1, Box 245D, Trafalgar, IN 46181.

DISPALURE, see Bait for Gypsy Moth

DOOM, see Milky Disease below, see also Bait for Japanese Beetle

DUCKS AND GEESE

A pair or a trio of ducks will help control the insect population if they can roam in fall and early spring. They will eat insects, but they will never scratch in the garden. Geese are said to eat weeds in strawberry beds and are widely employed in weeding large crops like cotton in the South. Ducks have been known to nip early spring onions pretty close to the ground, but they had earned them by also eating insects.

EGG SHELLS

Egg shells. Someone claimed that crushed egg shells would keep ants out of the pantry. This may differ with climates, or it may make a difference whether the egg shells are fresh or dried.

ELECTRONIC INSECT TRAP

Electronic insect traps are available from several sources. They attract insects to a light, where they touch a grid which electrocutes them. Some traps are equipped with a container into which insects fall as they are killed. Such a device is used to take a census of local insects. Available from Burpee Seeds, P.O. Box 6929, Philadelphia, PA 19132.

EQUISETUM TEA, see Bio-Dynamic Preparation 508

FISH OIL SOAP

Fish oil soap is an excellent wash for trees and plants where insects and eggs affect the bark. If it is smeared on trunks of trees, it will prevent worms from crawling up.

GLUE MIXTURE

A glue mixture may be used to imprison small insects such as aphis, red spider, scale insects, coccus and mealy bug. Dissolve ¼ pound of glue in 1 gallon of warm water overnight. A small quantity of flowers of sulfur may be added if necessary. Apply this with a sprayer to the twigs and leaves of the fruit trees without washing off the glue. When the mixture dries, it will flake off the twigs and leaves, and the imprisoned insects will also be lifted off. Several applications 7 to 10 days apart may be necessary to overcome all the insect generations in the middle of the summer season.

GYPTOL, see Bait for Gypsy Moth

HANDPICKING and LETTING THE INSECTS DECOMPOSE

Handpicking and letting the insects decompose to make Insect Repellent Spray: It is a general principle in nature that the waste of a species is distasteful to living members of that species: cows will not graze the grass around their droppings, rabbits are repelled by a solution of rabbit manure painted on tree trunks and so on. The same rule governs insects who are repelled by the decomposition of others of their own kind. If insects are collected and allowed to rot in a jar with a little water the resulting liquid will repel other insects of the same species. Some experiments have been made with electric insect traps to collect many different species to make one repellent for many kinds of insects.

Since the organs of smell in the insect are more sensitive than those of a human, the insects are repelled by a fine application of this bad-smelling substance. It is therefore possible to dilute the original material in water to make a very fine, dilute spray. In real practice, at least in the home garden, it is quite simple to leave the glass jar with the dead insects somewhere in the garden where it is effective as a repellent (see Rose Chafer). The only caution is to remove the glass jars when you expect fastidious guests to visit your garden.

HEMP SEED

Hemp seed to repel the Cabbage Butterfly. Loudon writes: "If in a patch of ground where cabbages are to be planted some hemp seed be sown all around the edge, in the spring, the strong smell which that plant gives in vapor, will prevent the butterfly from infesting the cabbages. The Russian peasantry, in those provinces where hemp is cultivated, have their cabbages within those fields by which they are free from caterpillars."

Although it is probably illegal to procure hemp seed at the present time because of its misuse in the form of a drug, the same principle might be tried to protect a garden area against the cabbage worm butterfly. Try sowing aromatic herbs around the outer borders. It might require several different herbs of perhaps different proportions because this is a different country, and both soil history and insect development may have changed since the above writing.

HORSERADISH TEA

Pour boiling water over horseradish leaves. Dilute this with four times as much water. Use for fungus on fruit trees at the beginning of an attack. Although fungus is not an insect, it is certainly a pest, and this is a recipe worth trying.

HOT WATER TREATMENT for HOUSE PLANTS

To expel aphis, mealy bug or scale from house plants, use clean hot water from 140°F, to 150°F. Cover soil in flowerpot with a crumpled newspaper or a circle of cloth with a slit. Put a handful of moss on top of the soil. While the plant is in a dormant state, tip the plant into a vessel of hot water for 5-10 minutes then set pot against two bricks to drain. Be sure it has plenty of air. A temperature of 140°F to 150°F will kill or expel any kind of mealybug, coccus, scale or other insect pests. It might be well to try this out on some cull plants before subjecting your favorite house plants to this rather rigorous treatment.

INSECT CONTROL BY TEMPERATURE

Most insects which infest houses can be killed by a temperature between 120°F and 130°F. It is possible in warm weather to turn on the heating system and to maintain this heat for about 6 hours, taking precautions, of course that no inflammable materials are too close to the heating system.

Few insects can stand a temperature of zero when they are brought out of a warm house. It is possible to kill moths in furniture or in clothing by taking the infested materials outdoors and leaving them out for a day or two in zero weather. This same principle is applied when we put corn meal or other food in the freezer to kill any insects that might have taken up residence in freshly ground flour.

KEROSENE TO KILL INSECTS

Touch any insect with a rag dipped in kerosene and that will be the end of it. Use a kerosene rag to kill egg clusters of gypsy moth or tent caterpillar or Brown-tail moth. The eggs will never hatch—alas!

A small amount of kerosene (about one-half cup) in a jar makes a lethal bath into which potato bugs, grasshoppers or other bad actors may be dropped to kill them. Be sure not to drop in the good ground beetles or preying mantis or lady bugs. Do not try to use this jarful of death and destruction for Insect Spray made from Hand Picked and Decomposed Insects.

KEROSENE EMULSION

Kerosene emulsion recipe for a small garden

2 cubic inches of soap dissolved in 1 pint of water, preferably rain water
1 quart of kerosene
Place in a bowl and beat with egg beater until a thick lathery cream.
Dilute this stock solution with 5 to 15 parts of water.
For dormant growth use 5 to 7 parts water;
for ordinary growth as it is in midsummer use 10 parts water;
for weak sprays use 15 parts water
Be sure to syringe or hose off plants with clean water after using

KIESELGUHR

Kieselguhr or Diatomaceous Earth also is called Tripoli and is used as a cleaning compound. This silicious substance is made up of the silicified skeletons of infinitesimal living creatures of the distant past. Made into a spray with water and sprayed over insects, it causes the death of today's insects. Scientists are still studying exactly what action these tiny silicious needles have on the bodies of small insects.

LADY BUGS

LADY BUGS, Bio-control Co., Rte. 2, Box 2397, Auburn, Calif. 95603.

LAVENDAR OIL

To drive fleas and sand fleas from a house in summer, sprinkle Oil of Lavendar on rugs and floors and turn heat on.

LIME AND WOOD ASHES

Mix hydrated or slaked lime with wood ashes and let stand a day or two. Pour from sprinkling can on squash plants and this will control the squash bug and its grey nymphs.

LIME TO CONTROL FLEA BEETLES

Loudon advises: "The sprinkling of slaked lime over the young plant is at once a safe and effectual process, and possesses the additional advantage of being beneficial to the plant. We are aware that it does not always succeed but we are inclined to attribute the failure to a bad quality of the lime, or a careless method of employing it. There should be enough put on to make the plants white, and they will be none the worse for the whitening. Dustings of ashes and soot are scarcely less effectual but . . . all such dustings should be done in the morning while the plants are still wet with dew. To apply a dusting at midday, when the sun shines gaily, is to waste time in merely amusing the beetles. A board newly painted white drawn over the plant on a still sunny day would soon become a black board by the attachment of myriads of flea beetles that would jump at it and remain upon it, the victims of their extravagant love of light. . . . This in common with all insects in the winged state, needs a dry air and some degree of warmth for its health and happiness. Many kinds of larvae need moisture, but no winged insect can abide long in moisture, and herein perhaps we may find a clue to the eradication of the flea beetle. By the simple process of *irrigating the plant* overhead three or four times a day, until the plant is out of the seed leaf and the danger is over, it is possible to *wash out* the flea beetle. . . ."

LIME TO KILL SLUGS AND SNAILS. See Slugs and Snails.

LIME WATER

Some of the oldest gardening books *(Forsythe On Trees)* recommend lime water to spray over fruit trees to control red spider. The recipe calls for water poured over the lime, the mixture allowed to stand until the lime has settled in the bottom. The clear liquid, then is sprayed over the trees.

METHYLATED SPIRIT, see, DENATURED ALCOHOL

MILK

MILK, some recent work seems to indicate that spraying milk over plants suffering from mildew or mould may help overcome fungus troubles.

MILKY DISEASE

Milky spore disease to combat Japanese Beetles has been isolated and is now manufactured and distributed commercially. The disease affects the insect in the grub stage so that it dies. It will not infect any other species than the Japanese beetle and it is approved by the USDA. This is available, called Doom, from Fairfax Biological Laboratory, Clinton Corners, New York. Many communities have purchased it to give to residents to overcome the beetle on a city-wide scale. See also Japanese Beetles.

MOTH BAITS, see Baits above, and *Other Pests*—

MOTH BALLS, to repel Rabbits, see *Rabbits*.

MULCH

Mulch is used for two purposes, to nourish and to protect. A mulch of compost or other plant food will nourish growing plants. For this purpose the compost does not need to be completely broken down, but as it does decompose it will be incorporated into the soil by cultivation. It may be added at any time of the growing season. If added during the very dry hot summer, it should be shaded with something to prevent its baking hard in the hot sun. A slight dusting of loose soil or a little shade from pulled-up weeds will be sufficient to save the goodness in the mulch material until it can settle into the garden topsoil.

The other kind of mulch is a protective mulch added to the top of the soil to keep the weeds down, to keep the soil moisture from evaporating, and to keep the ground surface protected from the sun and air. It is probably better not to leave such a mulch on the soil over winter, because it prevents the beneficial action of the frost in the earth. It also may harbor insects and mice. The mulch should not be left on over the winter if in any way it will hinder moisture from "coming and going" during the seasons of snow and ice. If a protective mulch is in such a condition that it will break down during the winter and become part of the topsoil, it may be left. But the home gardener should study this subject carefully and be sure he knows exactly what he is doing and why he is doing it.

MULCH, OAK LEAF

Mulch of Oak Leaves or tan bark produces a bitter atmosphere which slugs and cutworms and June bug grubs and other very tender ones just cannot stand. Perhaps these tender-bodied creatures just feel puckery all over when surrounded by bitter leaves like the oak.

MULCH, STINGING NETTLE

Gardeners who find the stinging nettle will not grow for them may still enjoy some of its benefits by importing stinging nettle plants from another garden to mulch certain special plants. It is even practical to bring "nettle soil" from the woods to scatter around plants in the backyard garden. The extra vitality which the nettle imparts to all its surroundings is remarkable, and instead of trying to get rid of it we should learn to appreciate it and to use it for its helpful qualities.

MULCH, TOBACCO STEM

The active quality in tobacco is present in tobacco stems which can be purchased from some seed houses. The stems are coarse enough to be easily handled, and their presence in a mulch will repel aphis, flea beetles and thrips. Their action may be furthered by dusting with powdered tobacco (see below) if necessary.

MUSTARD SEED FLOUR

Mustard Seed Flour has many different uses which as yet are not much publicized. Practical workers use it for various purposes. It will combat scale on fruit trees. When the mustard seed flour covers the scale insect, the latter suffocates. Use in the proportion of 1 pound of mustard seed flour to 10 gallons of liquid. Spray early in the spring. Spray with an oil spray in the fall for scale on fruit trees.

NAPHTHALENE FLAKES

Naphthalene Flakes may be scratched into the soil to repel insects which live just below the surface. Bear in mind, however, that soil organisms including the earthworms all are disturbed by any raw chemical substance which is added to the soil.

NASTURTIUM SEED

Nasturtium Seed may be sown around apple trees in spring to combat the woolly aphis. Sow a few nasturtiums in each hill of cucumbers to repel the cucumber beetles. Nasturtiums sown with broccoli will help repel aphis. In spite of the fact that growing nasturtiums often have a great many aphis of their own, in the process of growing with other plants, they seem to keep aphis away from their companion plants.

NETTLE, see Stinging Nettle.

NICOTINE

Nicotine is extracted from the tobacco plant. It is a highly toxic substance, especially in concentrated solutions. It is mixed with a soap solution to make it penetrate the waxy layer of the insects' skin, particularly aphis. Its residual toxicity lasts about one day to one week. Adding hydrated lime may make nicotine more potent. Do not use nicotine on roses. See also Tobacco, below.

NICOTINE SULPHATE

Nicotine Sulphate is a commercial derivative of the tobacco plant and is sold under several names. Usually it is about 40% active nicotine. It is very strong, (sometimes even its aroma will make a person sick), and should be used with extreme care and only in emergency situations. *Do not use on roses.* Hydrated lime may make Nicotine more potent. See also Tobacco, below.

ONION BREW

Recipes for an insect—repellent onion brew are as varied as the gardeners who concoct them. The brew should contain roots, stems and leaves—as many strong aromatics are possible: onion, garlic, horseradish, red pepper, mustard, garden mints. Chop fine or put in the blender with water. Add this pungent concentrate to a quart or more of water and add some liquid detergent. Then pour a generous amount over any plants that are infested with insects. If the brew ferments, it should work even better to repel insects.

PENNYROYAL HERB

Pennyroyal Herb rubbed fresh on the skin will repel mosquitoes for a short period of time. On kitchen shelves it repels ants.

PEPPERMINT HERB

Peppermint Herb has dozens of different uses. In stored grain it will keep rats out because: "Rats have so strong an aversion to the odor of peppermint they will not enter rooms or bins where it prevails."

PERSIAN INSECT POWDER, see Pyrethrum, below.

PLANTAIN

Plantain is an unpopular weed until people learn how valuable it is! It has such value as an emergency measure to stop bleeding that we try to have a few plants to use in case of need. Crush, or even bite the leaves to let out the juice and apply directly to the wound. Bleeding will stop, even from a deep cut. The tender inside leaves in early spring are a pleasant addition to a green salad. Plantain has been used for hundreds of years for healing broken bones! See "Romeo and Juliet," by William Shakespeare as the authority!

PYRETHRUM

Pyrethrum is also called Persian Insect Powder or Dalmatian powder. It is made of the ground up flowers of a relative of Feverfew or the Chrysanthemum Family. In order to be effective as an insecticide, the powdered flowers must be freshly ground. When an insect comes in contact with the active ingredient contained in pyrethrum, it is not killed but falls unconscious. For this reason, if pyrethrum spray is used on houseflies, they should be swept up and burned before they regain consciousness. Pyrethrum is not harmful to warm-blooded animals and people, except that some susceptible people may get a violent allergic reaction from it occasionally. Used with extreme care, it may be acceptable in a serious situation. Here are two recipes.

Pyrethrum Fly Spray

1 pound freshly ground pyrethrum flowers
1 gallon white kerosene
Mix and allow mixture to stand several hours. Shake

to extract the active principles from the pyrethrum. After sediment has settled, the clear liquid may be siphoned off and used for spraying flies.

Pyrethrum Spray for Insects on Plants

1 teaspoon to 1 tablespoon of fresh pyrethrum powder
2 quarts to 1 gallon of hot water
Add a little soft soap and let the mixture stand for awhile before using.

Each of these should be sprayed in a fine mist. Do not use the Fly Spray Recipe on plants because the kerosene will kill the plants. Pyrethrum is not effective if the solution is alkaline. Pyrethrum may also be used as a dust.

QUASSIA

Quassia comes in the form of ground-up wood chips which have an extremely bitter taste. It is especially effective for aphis, but it takes a long time to prepare it. For spray for a large operation the following recipe is:

Soak 1 pound quassia chips for 2 or 3 days in 8 gallons of water.
Simmer 2 or 3 hours over a slow fire to extract the bitter quality.
Strain. Mix with 1 pound of soft soap and churn together thoroughly.

For a small garden reduce the quantity to 2 ounces quassia in 1 gallon of water, but soak for the same length of time and simmer 2 or 3 hours also. Add 2 ounces of soft soap.

Another use for Quassia is in its extracted form. After soaking and simmering to extract its active principle, mix with Wormwood Tea (see below) to combat flying pests.

RADISH SEED

Loudon writes: "Cauliflower plants when first planted out, are frequently infested with flies or their larvae, to attract which, it is not uncommon to sow a little radish seed on the cauliflower ground a fortnight before transplanting; the flies preferring the tender leaves of the radish to those of the cauliflower, the latter are thus suffered to escape." Radish seed sown around each hill of cucurbits will repel the striped cucumber beetle until the plants get a good start.

RAGWORT

Ragwort, will prevent houseflies, a magazine article tells us. A sprig of ragwort hanging outside a doorway will drive houseflies away during the summer. The hardest thing is to be sure of your ragwort. Its botanical name is *Senecio Jacobeae* and it belongs to the family of *Composi-*

tae. It has feathered leaves and clusters of bright yellow-orange daisy-like flowers, each less than one inch across. The whole plant has a strong, pungent smell. It is not eaten by animals but is good for baby chicks.

RED PEPPER

Red pepper is a real help to the bee keeper in repelling ants from the bee hives. A liberal sprinkling of red pepper on kitchen shelf will send ants away in a hurry. It may also be helpful in keeping dogs from visiting evergreens every morning. Notice that *there is a difference* between Red Pepper, which is the very hot kind, and Cayenne Pepper, which is reddish in color but nowhere near as hot in taste. If you thought you tried red pepper and it didn't work, look again.

ROTENONE. see Derris.

RYANIA

Ryania (available Hopkins Agricultural Chemical Co. P.O. Box 584, Madison Wis. 53701) is a shrub from Trinidad which is gathered and shipped to America where it is ground into a powder used in some insecticides. It is specifically directed against the codling moth and the corn borer. This is a particularly useful material for a spray because it does not leave any toxic residues on the crops. There are constant changes going on in the field of toxic and non-toxic spray materials to keep insects under control on a large scale, and scientists are working steadily with their practical helpers to perfect these helpful sprays for saving our crops.

SALT

One way to deal with slugs is to sprinkle them with salt, but this should be done with the greatest care because even a little salt in the soil is injurious to plants. Salt in a weak solution is safe to use on cabbages, cauliflower and broccoli to destroy cabbage worms, but care should be used not to let much of the solution get into the soil. The reason salt is so effective is that it draws the liquids, through the process of osmosis, out of the insect's body and it cannot then survive.

SASSAFRAS OIL see Baits, for Fruit Tree Moth.

SOAP SUDS

When preparing to wash plants with soap suds to overcome aphis, coccus, scale insects or mealy bugs, it is best to use a neutral soap which will make thick suds. It is probably better for the health of the plant not to use strong water softeners or detergents because we still do not know what side effects these substances have on tender living tissues. Rainwater is naturally soft and will make a thick suds.

SOFT SOAP

Many garden sprays are concocted with soft soap as their basic ingredient. However, in the twentieth century one has to explain what soft soap is. This is a liquified or boiled solution of ordinary soap in water which becomes a jelly, which may then be further diluted to the desired consistency. Tincture of green soap is also practical to use in an insecticidal spray in fine dilution. Use only enough soft soap or green soap to whip into a rich suds to syringe or spray onto the plants. After drenching with soap suds, plants should always be rinsed off with clear water.

SOOT

Someone has suggested that we also need to explain what soot is, for a generation which has little experience of wood or coal fires and open chimneys. Soot is the black deposit which collects on the inside of a chimney and which should be removed at least once a year to prevent a chimney fire. It can be scraped from the flue, and it has many uses in the garden. See *Gooseberry caterpillar.* Contrary to the experience of that Head Gardener, another expert recommends that soot should be kept dry and not be used until it has been stored for four months.

SOOT AND WOOD ASHES to kill caterpillars. See gooseberry caterpillars.

STINGING NETTLE BREW

Place Stinging Nettle plants in a barrel or keg which has been sunk up to within 6 inches of the top in the ground. Cover the nettle plants with water and cover the barrel so that nothing can fall into it. In 5 to 7 days the nettles will have rotted and the liquid is ready for use. It will last a month or so and will be very green.

This Brew has a great many uses in the backyard garden, and the gardener who has watched closely as his plants have developed will know how and when to use it—on the plant or on the soil, or to help some flagging plant make use of the dynamic forces surrounding it. As an experienced elderly farmer once told us: "It gives the plant intelligence to know what to take from the atmosphere." Stinging Nettle Brew should be diluted with 6 parts of water when sprayed for green aphis. Spray several times if necessary. For black or gray aphis, dilute with 5 parts of water and spray as many times as necessary. When using this liquid manure as a fertilizer in the garden, be sure to dilute it so that it will not burn the leaves. Some gardeners use it directly on the soil. It may also be combined with equisetum tea for fighting pests. In fact, stinging nettle is so valuable and so versatile that it is used by the skillful gardener all the season from early spring greens for the table to stinging nettle hay to pack around apples and tomatoes in the storage cellar because of the preserving qualities of the nettle. Something to notice with one's own observation is the greater degree of vitality in plants growing near the stinging nettles in the woods. And this kind of observation can be put to good use in the garden if one works at it with patience and devotion.

SULFUR

Sulfur a specific remedy frequently used for mildew is dry flowers of sulfur. This may also be injurious if used injudiciously or on sensitive plants which cannot stand sulfur. It is safer to use Equisetum Tea (see Bio-Dynamic Preparation 508).

SWEET FERN. See Ants.

TANSY

Spread fresh tansy leaves on kitchenshelves to drive the ants back where they came from. It is also supposed to keep houseflies away. As a moth repellent it may be planted near peach trees to keep away flying insects which are especially fond of peaches—as people are too—because they are tender and delicious. Tansy, fresh or dried is a practical moth repellent for furs or woolens stored in the house. Rub dried tansy leaves into a dog's hair and his fleas will depart to happier environs. Fortunately for an herb which has so many uses, (people used to make a tansy pudding), it is extremely vital and can be transplanted anywhere at almost any time, and then can be trusted to spread and spread—even the sophisticated "non- spreading" varieties bought from a plant nursery.

TAR

The aroma given off by Tar is offensive to many insects, perhaps because it belongs to the old-time vegetation which has become extinct on this earth. See under Ant a description of the use of Tar to drive ants from a closet. This same principle might be used to advantage to drive out other insects. "To protect trees from mice: Tar 1 part, Tallow 3 parts. Melt the tallow and tar and mix them Apply melted with a paint brush around the trunk of the tree, preferably on a strip of cloth or paper so as not to injure the bark."

"Household Hints" (17th Century): "To overcome Potato Beetles . . Put a gallon of tar in a tub. Pour over it boiling water. Allow to settle and to cool. Sprinkle on Potato Vines." We rather suspect that the potato beetles referred to here are the little speckled kind which are now less of a threat than the Colorado Potato Beetle. Perhaps Tar would finish off the latter also.

TENT CATERPILLAR

The Tent Caterpillar is an avid American insect, and the Wild Cherry is an American tree. The tent caterpillar eats wild cherry and can always be found on these trees if there are any around. There is a principle which entymologists have discovered: If an insect is removed from its native feeding plant and learns to eat another plant, it will never return to its original feeding plant. In practical terms this means that if the Wild Cherry trees are destroyed, the tent caterpillar will go to other trees, principally, (you guessed it) the *apples!* And worse than that, it will never go back to the wild cherry. In other words *don't destroy the wild cherry trees.* They are valuable in that they concentrate the tent caterpillars where they can do little harm. Even though the cherry is defoliated, in about three weeks it may be in full leaf again. And we still have Calosoma Beetles and Braconid Wasps and other friends to help keep the tent caterpillars under control.

TOBACCO DUST

Tobacco Dust can be purchased from many seed houses. The dust is especially easy to use and effective to control aphis on chrysanthemums or other plants whose tender top growth is weakened by a heavy infestation of pests like aphis. Place a handful of tobacco dust in the center of a loosely woven cloth like cheese cloth, hold up the four corners of the cloth, and shake the dust over the aphis-laden plant. The fine dust surrounds the bodies of the aphis and either smothers them or overcomes them with the narcotic qualities in the tobacco. Do not inhale the dust because it will almost smother you, too!

TOBACCO STEM MULCH, see Mulch, Tobacco Stem

TOBACCO SMOKE FUMIGATION

Tobacco Smoke Fumigation. In light cases of aphis infestation, or when aphis first appear on house plants late in the winter, or on plants which cannot stand much hard treatment, it is possible to keep aphis and thrips under some control by blowing cigarette smoke through the plant every day. This is a rather precarious degree of control, however, and if the daily routine is interrupted or if there is a change in the weather, the aphis can quickly get beyond this kind of gentle control.

TOBACCO WATER BREW

Tobacco Water Brew. If you can get tobacco stems or tobacco dust, it is possible to make tobacco water brew at home in small quantities. We have also used plug tobacco which can be bought in any grocery store. Pack stems or cut up plug tobacco in a container and pour boiling water over them, enough to cover. Let this stand for several hours. As the water cools, the special qualities of the tobacco are extracted. Pour off the brown liquid and dilute with 1 part of tobacco brew extract to 4 parts of water. Use this only in extreme cases—in a fine spray—as it will kill good insects and soil bacteria as well as the bad insects. Keep bottle tightly covered and in the refrigerator but with a large label stating that it is most unpalatable and not to be mistaken for iced coffee!

TOBACCO WATER and SOFT SOAP

Tobacco Water and Soft Soap

Dissolve 1 pound of soft soap in 2 gallons of water.
Add 1½ pints of strong tobacco water made by soaking one cut up plug of tobacco in water (See Tobacco Water Brew)
Dip tops of plants in this solution, or sponge it on the leaves
Caution: before the plants are dry, rinse with pure water to remove the mixture which might otherwise burn the foliage.

Do not use any kind of Tobacco on roses. It will turn them black.

TOMATO PLANT EXTRACT

Tomato Plant Extract. From *Household Hints:* "Stems and leaves of tomato are well boiled in water and when liquor is cold, syringe over plants attacked by Green Fly (aphis). This at once destroys black or green fly, caterpillars and such. It also leaves behind a peculiar odor which prevents insects from coming back again."

TRAP CROPS

Some other trap crops:
white geranium and grapevine for Japanese Beetle
egg plant for Potato Bug
radish plants for Flea Beetles and
radish seed for Carrot Root Maggot Fly.

TRAPS FOR INSECTS

Traps for other insects: see
Earwigs
Electronic Insect Traps
Japanese Beetle Traps
Houseflies
Cankerworm
Woodlice
Harlequin Bugs
Tent Caterpillars

Loudon wrote in 1824: "The earwig and beetle trap is often only a hollow cylinder, but from this, if not taken regularly, at certain seasons, the insects escape. A close box, with an inverted truncated cone of glass in the center as a hopper, is better, because when earwigs, beetles, woodlice or such insects enter, they cannot escape and may be drowned or scalded or suffered to die there. The common bait is crumbs of bread . . The wasp and fly trap is merely a bottle half full of water honied at the mouth to entice their entrance. Some assert that the plant *hoya carnosa* whilst in bloom will attract wasps, and all other insects from the fruit in the house in which it grows, and others that boiled carrots will have the same effect." Of course, this refers to grapes or peaches grown in the greenhouse. Would boiled carrots distract wasps from the grapevine or the pear tree outdoors? Worth trying as an experiment!

TREE PASTE

Tree Paste. The recipe for the original home-made tree paste was 1/3 sticky clay, 1/3 cow manure, 1/3 fine sand. Vary the proportions if necessary. Mix with water to make a sticky paste which can be painted on the trunk and heavier limbs with a whitewash brush. Dilute the mixture with a liquid to make it thin enough to sprinkle or squirt into the upper branches, using a brush or sprayer with nozzle specially adapted to let coarse material pass through. Sometimes we have added Bio-Dynamic Preparation 508 (Equisetum Tea) and whole nasturtium plants including roots, put through a food grinder, the latter to repel aphis. This mixture is practical for a few trees only in the home orchard. In recent years other materials have been added to the original tree paste and it is now commercially available. Some gardeners have added Waterglass (see below) to the original tree paste recipe with excellent results.

TURKEYS

Turkeys. We have been told again and again that turkeys are better than anything for keeping bugs down. They never scratch a garden, they never miss a bug, and they are big enough to devour dozens every day. Unfortunately we listened to the wrong advisers and have never tried turkeys, but still hope to some day. In the meantime we pass on the information that they are highly recommended.

TURPENTINE

Turpentine is made from a kind of pine tree and it therefore carries a strong aromatic flavor which many insects cannot endure. Turpentine itself is such a strong substance that it will kill vegetation if it touched the leaves, but its fumes can be used to good advantage to repel insects, and, in the case of enclosed areas, to kill some of the more tender species.

TURPENTINE and WOOD ASHES

Turpentine and Wood Ashes to control Squash Bugs

Pour 4 Tablespoons of turpentine into ½ bushel of wood ashes at night.

The following morning, sprinkle lightly around squash plants to control the squash bugs.

WATERGLASS Waterglass is a commercial material (sodium silicate, sometimes potassium silicate). It is used in England to help plants and fruit trees combat the usual foggy atmosphere. A 2% waterglass solution is added to Equisetum Tea to help combat fungus diseases. A waterglass spray is used on fruit trees. Sometimes waterglass is added to the clay-cow-manure tree paste for fruit trees. Further information should be obtained from the Bio-Dynamic Agricultural Association, (see last page). With some of these newly-tried materials there are as yet no rules, and the home gardener will have to experiment and observe local conditions, and weather and various combinations of spraying ingredients—and then make his own rules. As a matter of fact, the home garden in the last analysis really depends on the gardener himself, or his own observations, skills and experience.

WOOD ASHES Wood ashes sprinkled liberally around the stems of cabbages will control club root and cabbage maggot fly; around onions it will control maggots and around beets will control scab. Wood ashes and hydrated lime on squash bugs will finish them off. See Turpentine and Wood Ashes.

WORMWOOD TEA Wormwood Tea. Gather wormwood leaves before ten o'clock in the morning and while the blooming process is still in the leaves and not yet in the bloom, that is, early in the summer. Dry in shadow, not in full sunshine. Use either dry or fresh leaves. Cover plants with water and bring to the boiling point. Take off the fire. Dilute with four parts water. Stir ten minutes to mix thoroughly. Use immediately against slugs and tender bodied insects like aphis or crickets. Houseflies do not like wormwood. Use dried wormwood to repel fleas.

in conclusion

There is always someone to ask the question, "Well, why *not* kill off all the bugs and get rid of them once and for all?" Usually the question is not answered during that conversation for the simple reason that it would take too long to explain. And they don't really want to know the answer anyway. In the first place, it would be impossible for man to kill off all the insects, and in the second place, even if he did, he wouldn't enjoy living in an insectless world. From a very superficial point of view the bugs seem to be a nuisance to human beings, but studied from a deeper, more thorough angle, the world of nature would be incomplete without these living creatures which all have work to do.

Man's shortcoming is in the way he thinks about bugs. Some people don't like them because they bite and raise welts and make them itch, or they crawl over masses of putrefying material and disgust us, and there are still others that run about the kitchen dipping into the sugar bowl or into the cake box. People are inclined to see only one insect at a time and to see that insect only in its relation to man —in the kitchen or in the garbage pail or on one's flesh (which we'll admit is a pretty personal relationship).

Let's stand a little farther apart and look at the insect kingdom, leaving man entirely out of the picture. Let's start from the ground and look upward. The soil is packed down by rain and snow. It needs lifting, aerating, turning over to make plants grow better. The earthworm turns it. The maggots clean up decomposition and return it to fertilize the soil. Plant roots reach out in the earth and soil microorganisms form a link between soil and roots, but there are also insects down there taking their part in the great work of manufacturing stuff for plants to live on. The whole plant kingdom is beholden to these invisible organisms, and to the insects for their part in the work.

Someday scientists will know as much about the rhizosphere—the sphere of root growth as it extends around the earth—as they now know about the atmosphere and the ionosphere and all the continuing spheres which echo out into space and beyond. When we have learned about this area, all men and women may come to have more respect for ants and maggots and other crawling things in the root sphere.

A few inches further up is the stem. Just think how many kinds of stems there are: tough ones and tender ones, crisp stems and juicy ones—all the way from the very thinnest vine up to the trunk of the largest tree. Each has its favorite dwellers from the insect world visiting it. Humans are so unperceptive and so preoccupied with less subtle attractions that they barely notice the tree trunk has its borers, just as the rose stem has its sucking aphis. We hardly know why they are there. We think of them as nuisances, but perhaps they also are there to help in the over-all balance of the natural kingdom.

One can't begin to recall all the so-called insect pests which are to be found on plant leaves—from the big green tomato worm which masquerades as a slightly rolled tomato leaf, to the leaf-cutting bee which snips circular pieces out of leaves to make cells for her offspring to inhabit. Remember now, we are thinking about insects and their jobs entirely apart from their impact on us human beings and our economy. Just in this leafy element alone, we know there are leaf rollers and tiers and miners and hoppers, and if we could know all the different climates and continents we would be overwhelmed with the different things insects do to leaves around this planet—quite aside from the climbing cutworms that eat the leaves in my garden, drat them!

But some of these insects, like the cutworm, outgrow their crawling, earthbound state and go on to higher places. After their long sleep in the pupal stage, they are transformed into air-borne creatures of infinite variety and beauty. These are the ones man notices with less apprehension and dislike. People are willing to accept the butterflies which adorn the flower garden in summer sunshine. But how about the moths that also fly there in the night? There are butterflies which inhabit jungles in the tropics and at the same time, butterflies that fly at unbelievable heights over snow-covered mountains. Some of the night flying "moths" are not moths at all but caddice flies whose larvae fed on water insects they captured in nets stretched across the surface of a trout stream. How can man possibly learn all about these variations around our earth?

Dr. Rudolf Steiner in one of his lectures mentions that the butterfly is like an extension of the flower petals fluttering in the air above the earth. The butterfly lifts earth forces into the air and at the same time brings air influences down to the flower—and back to the earth. There is need for scientists with reverence in their hearts to delve into these mysteries.

As yet, human knowledge of the work being done by the insects is too meager. When insect circles come tangent to human circles, sometimes there is understanding and patient study, sometimes a bug bomb—and then man's scientific enrichment is set back a bit or perhaps is just pushed further ahead in time.

If for nothing else, we hope the insect world can be appreciated for its unbelievable variety of shapes and sizes and preferences and occupations. There are biters and suckers and crawlers and swoopers. There are builders and scavengers and parasites and those that make honey. Until man's economics or woman's housekeeping enter the picture, the insect kingdom is noble and charming and infinitely ingenious. It is our hope that people will learn more about bugs so they may also know wonder and appreciation and reverence for the quality of life which surrounds us everywhere on this earth.

This is just *part* of the reason why we would rather not kill 'em all off and get rid of them once and for all!

suggested additional reading

A good library is essential for the person or family raising and storing food. No one can remember all of the information this requires, and a good library will provide it, at your fingertips. New ideas, techniques and theories are always being put forth, and the best way to keep up with them all is to keep your library up to date. There are many good books available; here are some Garden Way books that are excellent choices.

Secrets of Companion Planting for Successful Gardening, by Louise Riotte. 225 pp., quality paperback, $5.95. hardback, $8.95. For bigger, more luscious crops. It works.

Cash from your Garden: Roadside Farm Stands, by David W. Lynch. 115 pp., quality paperback, $3.95. Turn a big garden into a profitable family business.

Vegetable Garden Handbook, by Roger Griffith. 120 pp., spiral bound, $3.95. Take it into your garden, for information and your own record book.

Down-to-Earth Vegetable Gardening Know-How, featuring Dick Raymond. 160 pp., 8½×11, heavily illustrated quality paperback, $5.95. The beginner and the veteran gardener both love this book's proven advice.

Let It Rot! by Stu Campbell. 152 pp., quality paperback, $3.95. Homemade fertilizers for a healthier garden.

The Mulch Book, by Stu Campbell. 144 pp., quality paperback, $4.95. hardback, $5.95. Save work. Grow better vegetables. Everything on mulches.

These books are available at your bookstore, or may be ordered directly from Garden Way Publishing, Dept. 171X Charlotte, Vermont 05445. If your order is less than $10, please add 75¢ postage and handling.

alphabetical list of insects

index of recipes and formulas